THE CITY & GUILDS TEXTBOOK

LEVEL 3 NVQ DIPLOMA IN
ELECTROTECHNICAL
TECHNOLOGY 2357

UNIT 309

About City & Guilds

City & Guilds is the UK's leading provider of vocational qualifications, offering over 500 awards across a wide range of industries, and progressing from entry level to the highest levels of professional achievement. With over 8500 centres in 100 countries, City & Guilds is recognised by employers worldwide for providing qualifications that offer proof of the skills they need to get the job done.

Equal opportunities

City & Guilds fully supports the principle of equal opportunities and we are committed to satisfying this in all our activities and published material. A copy of our equal opportunities policy statement is available on the City & Guilds website.

Copyright

First edition 2014.

Reprinted 2017.

ISBN 978-0-85193-281-1

Publisher: Charlie Evans

Development Editor: Hannah Cooper

Production Editor: Fiona Freel

Picture Research: Katherine Hodges

Project Management and Editorial Series Team: Vicky Butt, Anna Clark, Kay Coleman, Jo Kemp, Karen Hemingway, Jon Ingoldby, Caroline Low, Joan Miller, Shirley Wakley

Cover design by Select Typesetters Ltd

Text design by Design Deluxe, Bath

Indexed by Indexing Specialists (UK) Ltd

Illustrations by Saxon Graphics Ltd and Ann Paganuzzi

Typeset by Saxon Graphics Ltd, Derby

Printed in Croatia by Zrinski

British Library Cataloguing in Publication Data

A catalogue record is available from the British Library.

Publications

For information about or to order City & Guilds support materials, contact 0844 543 0000 or centresupport@cityandguilds.com. Calls to our 0844 numbers cost 5 pence per minute plus your telephone company's access charge. You can find more information about the materials we have available at www.cityandguilds.com/publications.

Every effort has been made to ensure that the information contained in this publication is true and correct at the time of going to press. However, City & Guilds' products and services are subject to continuous development and improvement and the right is reserved to change products and services from time to time. City & Guilds cannot accept liability for loss or damage arising from the use of information in this publication.

City & Guilds
1 Giltspur Street
London EC1A 9DD

www.cityandguilds.com

publishingfeedback@cityandguilds.com

THE CITY & GUILDS TEXTBOOK

LEVEL 3 NVQ DIPLOMA IN
ELECTROTECHNICAL
TECHNOLOGY 2357
UNIT 309

PAUL HARRIS AND TREVOR PICKARD

SERIES EDITOR: PETER TANNER

ACKNOWLEDGEMENTS

City & Guilds would like to thank sincerely the following:

For invaluable knowledge and expertise
Eur Ing Darrell Locke, *BEng (Hons)*, *CEng*, *MIEE*, *MIEEE*, The Institution of Engineering and Technology, Technical Reviewer

Richard Woodcock, City & Guilds, Technical Reviewer and Contributor

For supplying pictures for the front and back covers
Jules Selmes

For their help with photoshoots
Jules Selmes and Adam Giles (photographer and assistant); Andy Jeffery, Ben King and students from Oaklands College, St Albans; Jaylec Electrical; Andrew Hay-Ellis, James L Deans and the staff at Trade Skills 4 U, Crawley.

Picture credits
Every effort has been made to acknowledge all copyright holders as below and the publishers will, if notified, correct any errors in future editions.

Emily Cooper p156; **Kewtech** p39; **RSI Video Technologies** p168; **Shutterstock** p1, p63, p66, p67, p154, p155, p156, p160, p161, p165, p166, p167; **Staco Energy Products Co.** p106; **Titan Products Ltd.** p168.

Text permissions
For kind permission of text extracts:

Permission to reproduce extracts from BS 7671:2008 is granted by the Institution of Engineering and Technology (IET) and BSI Standards Limited (BSI). No other use of this material is permitted. Pages 134, 139.

BS 7671: 2008 Incorporating Amendment No 1: 2011 can be purchased in hardcopy format only from the IET website http://electrical.theiet.org/ and the BSI online shop: http://shop.bsigroup.com

From the authors
Paul Harris: I would like to thank my wife, Carol, and sons, George and Alfie, for their support and for putting up with me while writing some of the sections.

Trevor Pickard: I need to express my sincere thanks to my wife, Sue, whose command of the English language has always been better than mine and without whose help the job of the copyeditor would have been that much harder!

Peter Tanner: Many thanks to my wife, Gillian, and daughters, Rebecca and Lucy, for their incredible patience; also to Jim Brooker, my brother-in-law, for showing me the importance of carrying out the correct safe isolation procedure on more than one occasion!

CONTENTS

ABOUT THE AUTHORS

Eur Ing **Paul Harris** *BEng (Hons), CEng, FIHEEM, MIEE, MCIBSE*

I am a Chartered Electrical Engineer, having started my career as a JIB Indentured Apprentice in 1978.

Following completion of my apprenticeship I worked in a variety of electrical installation jobs. Throughout my apprenticeship and time as a craftsman, through my enthusiasm to learn, I have always questioned 'why?', 'how?' and 'what will happen if…?' I believe it is important not to separate theory and learning from work life, and to take pride as an individual in work, development and the profession.

I returned to college in 1989, followed by university, and obtained an HNC Electrical Engineering and later BEng (Hons) in Building Services Engineering by part-time study.

In my work life I have held a number of engineer posts including managing direct labour work forces and teams of professional engineers. I have been a part-time lecturer for CGLI and BTEC ONC electrical Installation and have also written technical articles for a variety of institutions.

I currently have my own consultancy where I provide design expertise in specialist areas and installations to a variety of clients and industry professionals. Alongside this I am a Medical Locations expert for Working Group 710, represent the UK at IEC and am a committee member of JPEL64, which is responsible for the production of BS 7671.

My belief held in 1978 remains unchanged – training and development, with its supporting books and documentation, is invaluable to individuals and employers alike and should be valued throughout your career.

Trevor Pickard *IEng, MIET*

I am an electrical engineering consultant and my interest in all things electrical started when I was quite young. I always had a battery powered model under construction or an electrical motor or some piece of electrical equipment in various stages of being taken apart to see how they operated. Looking back, some of my activities with mains electricity would certainly be considered as unacceptable today!

Upon leaving school in 1966 I commenced work with an electricity distribution company, Midlands Electricity Board (MEB) and after serving a student apprenticeship I held a series of engineering positions. I have never tired of my involvement with electrical engineering and was very fortunate to have had a varied and interesting career in the engineering department of Midlands Electricity and embraced its various changes through privatisation and subsequent acquisition. I held posts in Design, Safety, Production Engineering, as Production Manager of a large urban-based operational division, as General Manager of the Repair and Restoration department, and as General Manager of the Primary Network department (33kV–132kV).

My interest in electrical engineering has extended beyond the '9–5 job' and I have had the opportunity to become involved in the writing of standards in the domestic, European and international arena with BSI, CENELEC and IEC and have for many years lectured for the Institution of Engineering and Technology.

Electricity is with us in almost every aspect of modern life and for those who are just starting their career in this field I would say keep an open mind, be safety conscious in how you carry out your work and who knows where your studies will take you.

Peter Tanner *MIET, LCGI*

Series Editor

I started in the industry while still at school, chasing walls for my brother-in-law for a bit of pocket-money. This taught me quickly that if I took a career in the industry I needed to progress as fast as I could.

Jobs in the industry were few and far between when I left school so after a spell in the armed forces, I gained a place as a sponsored trainee on the CITB training scheme. I attended block release at Guildford Technical College where the CITB would find me work experience with 'local' employers. My first and only work experience placement was with a computer installation company located over twenty miles away so I had to cycle there every morning but I was desperate to learn and enjoyed my work.

Computer installations were very different in those days. Computers filled large rooms and needed massive armoured supply cables so the range of work I experienced was vast from data cabling, to all types of containment systems and low voltage systems.

In the second year of my apprenticeship I found employment with a company where most of my work centred around the London area. The work was varied, from lift systems in well-known high-rise buildings to lightning protection on the side of even higher ones!

On completion of my apprenticeship I worked for a short time as an intruder alarm installer mainly in domestic dwellings, a role where client relationships and handling information is very important.

Following this I began work with a company where I was involved in shop-fitting and restaurant and pub refurbishments. It wasn't long before I was managing jobs and gaining further qualifications through professional development. I was later seconded to the Property Services Agency designing major installations within some of the most well-known buildings in the UK.

A career-changing accident took me into teaching where I truly found the rewards the industry has to offer. Seeing young trainees maturing into qualified electricians is a worthwhile experience. On many occasions I see many of my old trainees when they attend further training and update courses. Seeing their successes makes it all worthwhile.

I have worked with City & Guilds for over twenty years and represent them on a variety of industry committees such as JPEL64, which is responsible for the production of BS 7671. I am passionate about using my vast experience in the industry to maintain the high standards the industry expects.

HOW TO USE THIS TEXTBOOK

Welcome to your City & Guilds Level 3 NVQ Diploma in Electrotechnical Technology textbook. It is designed to guide you through your 2357 Level 3 qualification and be a useful reference for you throughout your career.

It covers everything you will need to understand in order to complete your written or online tests and prepare for your practical assessments. Across some learning outcomes in the 2357 units there is some natural revisiting of knowledge and skills in different contexts, which the content in this book also reflects, for practical use and reference.

Throughout this textbook you will see the following features:

KEY POINT

An ammeter has a very low internal resistance and must never be connected across the supply.

KEY POINT These are particularly useful points that may assist you in revision for your tests or to help you remember something important.

Torque

A force that causes rotation.

DEFINITIONS Words in bold in the text are explained in the margin to aid your understanding. They also appear in the glossary at the back of the book.

ACTIVITY

Put these conductors in ascending order of resistivity: copper, steel, aluminium, brass, silver, gold.

ACTIVITY These provide questions or suggested activities to help you learn and practise.

ASSESSMENT GUIDANCE

When working out transformer ratios, use the winding to winding values – that is, phase voltage to phase voltage (Vp-Vs).

ASSESSMENT GUIDANCE These highlight useful points that are helpful for your learning and assessment.

 SmartScreen Unit 309
Handout 13

SMARTSCREEN These provide references to City & Guilds online learner and tutor resources, which you can access on SmartScreen.co.uk.

Assessment criteria

3.2 Explain the principles of basic mechanics as they apply to levers, gears and pulleys

ASSESSMENT CRITERIA These highlight the assessment criteria coverage through each unit, so you can easily link your learning to what you need to know or do for each Learning outcome.

Where tables and forms in this book have been used directly from other publications such as the IET this has been noted, and the style reflects the original (with kind permission). Always make sure you use the latest information and forms.

You will also see the following abbreviation in the running heads:

LO – learning outcome **LO6** **Electrical supply and distribution systems**

Understanding the electrical principles associated with the design, building, installation and maintenance of electrical equipment and systems

This unit provides an introduction to the basic electrical principles. An understanding of these principles provides the theory behind the practical requirements when designers undertake an installation project and maintenance teams carry out both planned and emergency maintenance.

LEARNING OUTCOMES

There are twelve learning outcomes to this unit. The learner will:

1 Understand mathematical principles which are appropriate to electrical installation, maintenance and design work

2 Understand standard units of measurement used in electrical installation, maintenance and design work

3 Understand basic mechanics and the relationship between force, work, energy and power

4 Understand the relationship between resistance, resistivity, voltage, current and power

5 Understand the fundamental principles which underpin the relationship between magnetism and electricity

6 Understand electrical supply and distribution systems

7 Understand how different electrical properties can affect electrical circuits, systems and equipment

8 Understand the operating principles and applications of d.c. machines and a.c. motors

9 Understand the operating principles of different electrical components

10 Understand the principles and applications of electrical lighting systems

11 Understand the principles and applications of electrical heating

12 Understand the types, applications and limitations of electronic components in electrotechnical systems and equipment.

The unit will be assessed by:

■ Assessment A – closed book online e-volve multiple-choice assessment, covering learning outcomes 1–4 of this unit.

■ Assessment B – closed book short-answer assessment covering learning outcomes 5–12 of this unit.

Assessment criteria

1.1 Identify and apply appropriate mathematical principles which are relevant to electrotechnical work tasks

ASSESSMENT GUIDANCE

Any number which is raised to the power of zero (0) has a value of 1, so $10^0 = 1$ and $25^0 = 1$ and $3.142^0 = 1$ and so on.

SmartScreen Unit 309

PowerPoint 1 and Handout 1

INDICES

Indices are used to replace repetitive multiplications. For example, $10 \times 10 \times 10 = 1000$, so the calculation could be written easily by saying 10^3, which means ten multiplied by itself twice, or three lots of ten multiplied together.

Where indices are negative, the value becomes a fraction. For example:

$$5^{-1} = \frac{1}{5}$$

$$\text{or } 5^{-2} = \frac{1}{25}$$

$$\text{or } 5^{-3} = \frac{1}{125}$$

Most calculators will have a (x^2) button to square a number and scientific calculators also have a button (x^y), which allows a number to be raised to any power or index. For example, to calculate 5^5, use buttons **5** x^y **5** = 3125. This is much easier than keying $5 \times 5 \times 5 \times 5 \times 5$.

Generally, in electrical science and principles, large values are used, such as thousands of watts or millions of ohms. Other aspects of electrical work deal with tiny amounts, such as millionths of an ampere or thousandths of an ohm. This can become a problem in calculations, as errors may occur if the correct number of zeros is not entered into the calculator. Instead of inserting the actual number with lots of zeros, we use 'to the power of ten'.

The 'power of' numbers are given names that are explained in the table opposite. There is less chance making an error using this method.

Numbers expressed as indices (to the power of 10)

Actual number	Number shown to the power of 10	Prefix used
1 000 000 000 000	10^{12}	tera (T)
1 000 000 000	10^9	giga (G)
1 000 000	10^6	mega (M)
1000	10^3	kilo (k)
100	10^2	hecto (h)
10	10^1	deka (da)
0.1	10^{-1}	deci (d)
0.01	10^{-2}	centi (c)
0.001	10^{-3}	milli (m)
0.000 001	10^{-6}	micro (μ)
0.000 000 001	10^{-9}	nano (n)
0.000 000 000 001	10^{-12}	pico (p)

ASSESSMENT GUIDANCE

When you are writing down 'values of' in milli or mega units, always make sure that mega is a capital M and milli is a lower case m. When your work is marked, it has to be clear what unit you are using.

To perform this on a calculator, use the button marked **EXP** (it may also be marked **×10**) then insert the index (the 'to the power of' number).

To perform a complex calculation such as 325 giga × 5 micro ÷ 12 mega, you would need to insert a lot of zeros before and after the decimal point, like this:

$$\frac{325\,000\,000\,000 \times 0.000\,005}{12\,000\,000}$$

To make it easier to perform on a calculator, use the indices and the **EXP** button, so the formula becomes:

$$\frac{325 \times 10^9 \times 5 \times 10^{-6}}{12 \times 10^6}$$

So, on a calculator **325** **EXP** **9** **×** **5** **EXP** **–6** **÷** **12** **EXP** **6** **=** will give the result 0.14.

If you try calculating $25 \times 10^6 \times 2 \times 10^3$, depending on your calculator, the result obtained may be 5 with a 10 in the right of the screen. If you press the button marked **ENG** or **SHIFT ENG**, you will see the 'to the power of' number change to 50 to the power of 9, which is equal to 50 G, or, if you keep pressing the **ENG** button, you may eventually get 50 000 000 000. All the numbers displayed have the same values, just represented in different ways.

KEY POINT

Calculators do not all work in the same way. Therefore, any calculator key sequence suggested is only an example of what would be required on a standard scientific (non-programmable) calculator. If the key sequence does not give the expected result on your calculator, either ask your tutor for advice or refer to the manual for your calculator.

KEY POINT

Be wary as sometimes the result on a calculator can be expressed as a 'to the power of', so keep an eye on any numbers to the right of the result in the calculator screen.

TRANSPOSING BASIC FORMULAE

Algebra

The branch of mathematics that uses letters and symbols to represent numbers, to express rules and formulae in general terms.

In electrical science we use **algebra** all the time when dealing with formulae. For example, Ohm's law states that a voltage can be determined by multiplying current by resistance. Instead of writing it out in full, we use letters and symbols to represent unknown values, which are known as variables. Therefore $V = I \times R$ is a basic use of algebra. Algebra can be used to show relationships between different quantities.

If, for example, we want to find the total cost, b, of four bars of chocolate, and the price of one bar is a, the formula would be:

$$b = 4a$$

Many formulae are used in electrical science. It is much easier to remember one particular formula in one particular way. For example:

$$R = \frac{\rho l}{A}$$

where:
R = resistance, in ohms (Ω)
ρ = resistivity value of a particular material (Ωm)
l = length of a cable conductor
A = cross-sectional area of the conductor.
(This formula is explained in more detail on page 40.)

Transposition

Rearranging a formula to make the unknown you need to find, the subject of the formula.

The above formula could be rearranged, to determine the formula for A (see page 6). This is called **transposition**.

To learn the rules of transposition, you need to consider three types of formula:

- those that use addition and subtraction
- those that use multiplication and division
- those involving both (mixed formulae).

The following methods follow simplified mathematical rules and will give you the ability to transpose any formulae you come across at Levels 2 and 3.

Transposing formulae involving addition and subtraction

The rules of transposition for addition and subtraction

1 The unknown you want to find must be on its own on one side of the equals sign.

2 The unknown should not have a minus sign in front of it.

3 Any unknown that moves over the equals sign has the sign in front of it changed from addition to subtraction or from subtraction to addition.

4 Any unknown that does not have a sign (positive or negative) in front of it is assumed to be positive.

Consider how to transpose this formula to find c.

$$a + b + c - d = e$$

The unknown c has a plus sign (+) in front of it so it stays where it is and the other unknowns around it need to move. The unknowns a and e have no sign in front of them so assume they are not being subtracted from anything.

So, to transpose it:

$a + b + c - d = e$	move d over the equals sign
$a + b + c = e + d$	$-d$ changes to $+d$ when moved
$a + c = e + d - b$	move b, changing the + to −
$c = e + d - b - a$	move a (changing the sign) to leave c alone
$c = e + d - b - a$	the finished, transposed formula.

ACTIVITY

There are many formulae that will have to be transposed as you progress through the course. Get as much practice as you can. Start with all the formulae you can think of with three items, then four and so on.

At all times, the equal sign acts as the centre of a set of balance scales and the formula remains balanced. If numbers are substituted for the letters, this might give:

$10 + 20 + 15 - 5 = 40$	move numbers away from 15 over the equals sign
$10 + 20 + 15 = 40 + 5$	see that the formula is balanced as it equals 45 on each side
$10 + 15 = 40 + 5 - 20$	keep the balance and move the 20
$15 = 40 + 5 - 20 - 10$	move 10 to leave 15 alone
$15 = 40 + 5 - 20 - 10$	the finished, transposed formula.

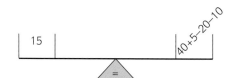

Formulae involving addition and subtraction remain balanced over the equals sign.

Transposing formulae involving multiplication and division

Formulae involving division often include parts that are made up of letters and numbers written above and below a line, as fractions. The number above the line is called the numerator, and the number below is the denominator. A denominator is the number of parts a whole is divided into, a numerator is the number of these parts that we are dealing with. So, thinking of ¾ of a cake, the cake is divided into four equal parts, we have three of them.

$$\frac{\text{numerator}}{\text{denominator}} = \frac{\text{number of parts we have}}{\text{number of equal parts in the whole}}$$

The rules of transposition for multiplication and division

1 The unknown you want to find must be on its own on one side of the equals sign.

2 The unknown to be found must be on its own, not part of a fraction.

3 Any unknown moved over the equal sign changes from top to bottom or bottom to top.

4 Any unknown or number can be written as a fraction by writing it over a denominator of 1, for example, $4 = \frac{4}{1}$.

The following formula can be transposed, to find a formula for A:

$$R = \frac{\rho l}{A}$$

The A is at the bottom and it needs to be at the top. This is done by moving it over the equals sign. The letter R can be written as a fraction, as $\frac{R}{1}$. So moving A over the equals sign gives:

$$\frac{RA}{1} = \frac{\rho l}{1}$$

Now all the unknowns are at the top. Remember that anything divided or multiplied by 1 remains the same, but writing the 1s in helps us to remember that there is a top and bottom. Now divide both sides by R.

$$\frac{A}{1} = \frac{\rho l}{R}$$

R has now been moved over the equals sign to leave A alone, so:

$$A = \frac{\rho l}{R}$$

This can once again be demonstrated by using numbers to see how balance is maintained:

$$50 = \frac{20 \times 5}{2}$$

Insert a 1 to show 50 as a fraction:

$$\frac{50}{1} = \frac{20 \times 5}{2}$$

Move the 2 over the equals sign from bottom to top:

$$\frac{50 \times 2}{1} = \frac{20 \times 5}{1}$$

To keep the balance, move the 50 over the equals sign from top to bottom:

$$\frac{2}{1} = \frac{20 \times 5}{50}$$

ACTIVITY

Transpose each of the following formulae to find the unknown indicated.

$E = V - I_a R_a$ Find V.

$E = \phi N$ Find N.

$P = I^2 R$ Find I.

$E = \frac{2P\phi NZ}{A}$ Find N.

$P = V \times I \times \cos\theta$ Find $\cos\theta$.

Then to finish off, remove the 1 from the left-hand side to get:

$$2 = \frac{20 \times 5}{50}$$

Job done!

With practice, this routine becomes second nature. You just need to remember the rules.

Transposing mixed formulae

This requires combining all the rules. Sometimes numbers or unknowns need to be combined, so they are grouped in brackets. This effectively makes each group act as if it is a single number or unknown. For example, to determine the unknown *d* from the following formula:

$$\frac{(a + b) \times c \times d}{e} = f$$

As *d* is at the top, it needs to be left where it is and the other unknowns are moved. As the (*a* + *b*) is in brackets, the whole thing can be moved together and treated as a single unknown. Remember, *f* is also over 1.

$$\frac{(a + b) \times c \times d}{e} = f$$

Move (*a* + *b*) over the equals sign, so:

$$\frac{c \times d}{e} = \frac{f}{(a + b)}$$

And:

$$\frac{d}{e} = \frac{f}{(a + b) \times c}$$

So finally:

$$\frac{d}{1} = \frac{f \times e}{(a + b) \times c}$$

Or:

$$d = \frac{f \times e}{(a + b) \times c}$$

Once again, to prove this with numbers:

$$\frac{(3 + 2) \times 4 \times 10}{25} = 8$$

The (3 + 2) is treated as a single number (ie 5), so:

$$\frac{4 \times 10}{25} = \frac{8}{(3 + 2)}$$

Formulae involving multiplication and division remain balanced over the equals sign

ASSESSMENT GUIDANCE

Make sure you practise using as many formulae as you can. Make up your own if you wish and test them out by substituting numbers.

Then:

$$\frac{10}{25} = \frac{8}{(3+2) \times 4}$$

And finally:

$$10 = \frac{8 \times 25}{(3+2) \times 4}$$

There is one further rule to remember:

Whatever you do to one side of the formula, you must keep the balance on the other.

To transpose or rearrange the formula below to make b the subject:

$$\sqrt{a^2 + b^2} = c$$

the square root must be eliminated because it locks in b. The opposite of taking a square root is to square, so:

$$a^2 + b^2 = c^2$$

Applying the rules gives:

$$+b^2 = c^2 - a^2$$

As the unknown needed is b, not b^2, take the square root on both sides:

$$b = \sqrt{c^2 - a^2}$$

Triangles and Pythagoras' theorem

Triangles are used to quantify electrical values. Later you will explore power triangles and phasor diagrams as a way of determining circuit values. To help with these, you need to understand basic principles of trigonometry (see page 9) and Pythagoras' theorem.

The triangle below is a right-angled triangle. Given the lengths of any two sides of a right-angled triangle, you can use Pythagoras' theorem to find the length of the third side. The **hypotenuse** of a right-angled triangle is always opposite the right angle.

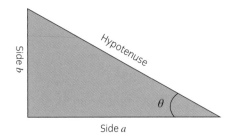

A right-angled triangle

Pythagoras discovered that the square of the length of the hypotenuse is equal to the square of the length of side a added to the square of the length of side b. Or, to express this as a formula:

$$a^2 + b^2 = h^2$$

or:

$$\sqrt{a^2 + b^2} = h$$

The length of the hypotenuse for the triangle below can therefore be calculated by applying Pythagoras' theorem.

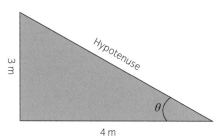

Applying Pythagoras' theorem

As:

$$\sqrt{a^2 + b^2} = h$$

Then:

$$\sqrt{3^2 + 4^2} = 5 \text{ (the length of the hypotenuse)}$$

ASSESSMENT GUIDANCE

You will find calculations based on triangles and Pythagoras all through electrical science work. This will include power triangles, lighting calculations and power factor correction amongst others.

Trigonometry

Trigonometry is used extensively in engineering and construction technology as well as many other sciences. Without trigonometry we would not be able to establish the heights of hills, mountains and buildings or the distance to stars and other planets.

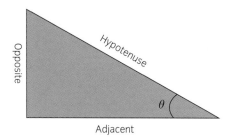

The three sides of a right-angled triangle are: adjacent to the angle theta (θ), opposite the angle theta (θ) and the hypotenuse, which is opposite the right angle.

Consider this explanation to help you understand the relationships in trigonometry.

Looking at a right-angled triangle, if any of the sides increase or decrease in length, the angle θ changes accordingly. (Unknown angles are given the symbol theta, θ, from the Greek alphabet.)

There are three formulae to remember when it comes to trigonometry.

$$\sin \theta = \frac{\text{opposite}}{\text{hypotenuse}}$$

$$\cos \theta = \frac{\text{adjacent}}{\text{hypotenuse}}$$

$$\tan \theta = \frac{\text{opposite}}{\text{adjacent}}$$

Some people remember them by the mnemonic SOH CAH TOA.

The sine, cosine and tangent ratios for the angle are calculated by using the relationships between the lengths of the sides. For example, the triangle shown here has the following dimensions:

- adjacent = 3 m
- opposite = 4 m
- hypotenuse = 5 m.

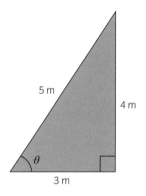

ASSESSMENT GUIDANCE

The internal angles of a right-angled triangle always add up to 180°.

Using these values you can work out the sine, cosine and tangent values of the angle.

$$\sin \theta = \frac{\text{opposite}}{\text{hypotenuse}} = \frac{4}{5} = 0.8$$

$$\cos \theta = \frac{\text{adjacent}}{\text{hypotenuse}} = \frac{3}{5} = 0.6$$

$$\tan \theta = \frac{\text{opposite}}{\text{adjacent}} = \frac{4}{3} = 1.333$$

With these ratios, the angle θ can be found using a calculator:

- $\sin^{-1} 0.8 = 53.1°$
- $\cos^{-1} 0.6 = 53.1°$
- $\tan^{-1} 1.333 = 53.1°$

Each of the ratios determined by using the different lengths relates to the same angle. We can use these different ratios to determine any missing value from the triangle.

You need to select the right formula to do this. Choose the one that uses the information you already have and that gives you what you need to find. For example, if you knew the length of the hypotenuse and the angle, but needed to determine the length of the opposite, you would choose the formula that uses all three. This is the sine formula.

Values of sine, cosine and tangent are ratios that have been calculated for every possible angle. Many years ago, before calculators, these factors were found from books of mathematical tables. These days, all of the different possible values are programmed into your scientific calculator.

When using the **SIN**, **COS** and **TAN** functions on a calculator, pressing each button directly will provide the sine, cosine or tangent value (ratio) for the given angle. For example, **SIN 45** will give a value of 0.7071, which means an angle of 45° has a sine value of 0.7071. To find a value of angle from a calculated or given sine, cosine or tangent, you will need to use the **SHIFT** or **2nd Function** button followed by the sin⁻¹, cos⁻¹ and tan⁻¹ symbols. With this book, if you see the function cos⁻¹, tan⁻¹ or sin⁻¹, the **SHIFT** or **2nd Function** button is required for that calcuation.

ACTIVITY

Sine, cosine and tangent are the three main trigonometric ratios. Identify three others.

KEY POINT

Calculators do not all work in the same way. Therefore, any calculator key sequence suggested is only an example of what would be required on a standard scientific (non-programmable) calculator. If the key sequence does not give the expected result on your calculator, either ask your tutor for advice or refer to the manual for your calculator. Ensure your calculator is set to DEG (for degrees), not RAD or GRAD or your results will be very different. Radians and gradians are other methods of measuring angles.

Sine

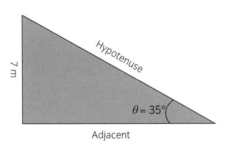

To determine the length of the hypotenuse of the triangle above, as you know the value of the angle and the length of the opposite side, use the trigonometric function sine as it includes the three known or needed ingredients.

Use the formula:

$$\sin \theta = \frac{o}{h}$$

Transpose the formula:

$$h = \frac{o}{\sin \theta}$$

so:

$$h = \frac{7}{\sin 35°} = 12.2 \text{ m}$$

Example

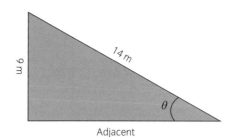

Use the sine function to determine the value of the required angle.

Then:

$$\sin \theta = \frac{o}{h}$$

so:

$$\sin \theta = \frac{9}{14} = 0.642$$

This is not the actual value of the angle, but the sin of the angle. In order to find the angle from this, use the $\sin^{-1}$ function by pressing the **SHIFT** button or second function button, depending on your calculator. So the angle is found as:

$$\sin^{-1} 0.642 = 40°$$

So on the calculator, press **SHIFT SIN 0.642 =**

The sequence **9 ÷ 14 = sin⁻¹ =** will give the angle as 40°. This answer is more accurate as the calculator remembers all of the values after the decimal point in the value of the sine of the angle.

KEY POINT

The button on your calculator marked **ANS** will insert the answer of the last calculation into the one being performed.

Cosine

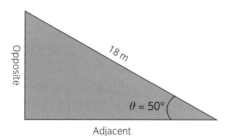

To determine the length of the side adjacent to θ in the triangle above, use cosine as you know the size of the angle and the length of the hypotenuse, and you need to find the length of the adjacent side.

Use the formula:

$$\cos \theta = \frac{a}{h}$$

Transpose the formula:

$$a = \cos \theta \times h$$

so:

$$a = \cos 50° \times 18 = 11.57 \text{ m}$$

Example

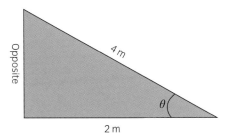

To determine the angle θ of the triangle above, use the formula:

$$\cos \theta = \frac{a}{h}$$

So:

$$\cos \theta = \frac{2}{4} = 0.5$$

Then $\cos^{-1} 0.5 = 60°$.

Tangent

Where calculations in trigonometry do not involve the hypotenuse of a triangle, use the tangent function.

Example

A situation requires the height of a building to be determined. With the use of a very simple theodolite made from a protractor, straw and a pole of known length, the height can be determined by measuring a distance from the building and measuring the angle, by looking through the straw, to the top of the building.

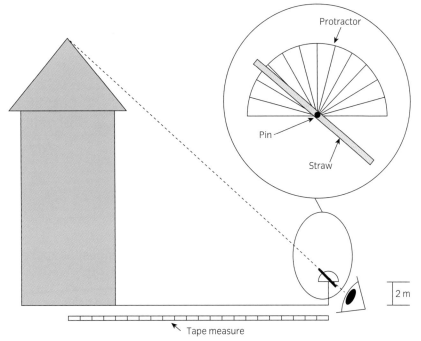

Using trigonometry to measure the height of a building

Assuming that the distance measured out is 20 m, the pole being used is 2 m long and the angle measured to the top of the building is 70°, the height of the building is found by using the formula:

$$\tan\theta = \frac{o}{a}$$

Transpose the formula:

$$\tan\theta \times a = o$$

Therefore, from the top of the pole, the height is:

$$\tan 70° \times 20 = 54.9 \text{ m}$$

The overall height (rounded up) is 57 m, once the pole length of 2 m is included.

Statistics

Statistics is the branch of mathematics that includes collecting, organising and interpreting data. In the electrotechnical sector, data may be used in situations such as:

- cable selection tables
- maximum permitted values of earth fault loop impedance
- quotations and tendering
- billing
- quantity surveying
- monitoring job progression.

Statistics are best summarised in table form and can be presented as graphs. Below is an example of how data may be presented and analysed.

Wholesaler	Cost of socket per unit (£)	Discount offered (%)	Cost after discount (£)
ABC Supplies	2.52	20	2.02
Sockets R US	2.45	8	2.25
Lecky World	3.65	30	2.55

From the data, it is clear that socket outlets are cheapest at Sockets R US but, once discount is applied, ABC Supplies is best.

Using computer spreadsheet programs, much of the analysis can be done quickly and automatically.

Percentages

A percentage (%) is a ratio expressed as a fraction of 100. For example, 35% is equal to $\frac{35}{100}$. A percentage is used to express one value or quantity as a fraction of another.

Percentages are widely used in everyday life to work out things such as how much tax somebody pays or to show how much discount you will get off material at a wholesaler.

For example, if you have to pay 20% of your earnings each month in tax, and you earn £1400.00 per month you can work out the amount of tax to pay by calculating:

$$\frac{1400 \times 20}{100} = 280 \text{ or } £280.00$$

A shorter way to do this is to show the percentage as a decimal multiplier, so 20% becomes $\frac{20}{100} = 0.2$.

so £1400.00 × 0.2 = £280.00.

If you have to pay an extra 20% VAT on goods, and the goods cost £28.00 excluding VAT, you need to add 20% of £28.00 to the cost, so

$\frac{28 \times 20}{100} = £5.60$. So you need to add £5.60 to the ex-VAT price of

£28.00, so the cost including VAT is £33.60.

Once again, a shorter way to do this is to use a decimal multiplier. This time we need to add 20% so we use a multiplier of 1.2. This essentially adds 20% to the initial cost as it multiplies the value by 1, keeping the £28, and adds a further 0.2 representing the 20% increase.

The use of percentages as multipliers is very common in the electrotechnical industry as they are often used to determine temperature changes and effects when designing electrical installations.

KEY POINT

Multipliers are often referred to as factors in publications such as BS 7671 and are commonly used in circuit design calculations.

ACTIVITY

A computer in a sale has a 30% discount. The full price is £429.99. What is the sale price?

Assessment criteria

2.1 Identify and use internationally recognised (SI) units of measurement for general variables

SmartScreen Unit 309

PowerPoint 2 and Handout 2

SI UNITS

As the world of science, and electrical science in particular, developed, it became necessary to agree on some form of standardisation so that scientists could understand one another's work and share their ideas.

The system of SI units (short for *Système International d'Unités*) is internationally recognised and based on the metric system. The alternative, based on the imperial system of measurement and still used in the USA (US imperial), is far more complex and less elegant, as the non-electrical comparisons below between SI and imperial demonstrate.

Imperial

1 ton = 20 cwt (hundred-weight) = 2240 lb (pounds weight)
= 35 840 oz (ounces)

Metric

1 tonne = 1000 kg (kilograms) = 1 000 000 g (grams)

There are seven base SI units, which then generate many derived units.

Base SI units

Quantity	Quantity symbol	Unit name	Unit symbol
Current	I	ampere	A
Length	l	metre	m
Luminous intensity	I	candela	cd
Mass	m	kilogram	kg
Temperature	T	kelvin	K
Time	t	second	s

> **KEY POINT**
>
> There are seven base SI units in total. The seventh unit is the measure of substance which is the mole. This unit is mainly used in chemistry to measure chemical substances. It is not really relevant to Electrotech.

ELECTRICAL SI UNITS

In electrical applications, units of measurement are standardised. You must always ensure that any calculation or formula uses the base SI unit or the derived units shown below. If a calculation uses a different unit, the result will be incorrect.

Standardisation of units of measurement is essential to be sure that everybody knows the correct measurements to use. There have been many instances over the years where very complex and large international projects have gone disastrously wrong because Europe uses metric units and America uses imperial.

The following table lists the most commonly used electrical units of measurement:

Most commonly used electrical units of measurement

Quantity	Quantity symbol	Unit name	Unit symbol
Area	A	square metre	m^2
Capacitance	C	farad	F
Charge	Q	coulomb	C
Energy (work)	W	joule	J
Force	F	newton	N
Frequency	f	hertz	Hz
Impedance	Z	ohm	Ω
Inductance	L	henry	H
Magnetic flux	Φ	weber	Wb
Magnetic flux density	B	tesla	T
Potential difference	V	volt	V
Power	P	watt	W
Reactance	X	ohm	Ω
Resistance	R	ohm	Ω
Resistivity	ρ	ohm-metre	Ωm

Assessment criteria

2.2 Identify and determine values of basic SI units which apply specifically to electrical variables

ASSESSMENT GUIDANCE

Energy is normally measured in joules (J), but for many purposes this unit is too small so the kilowatt hour (kWh) is used. A 1 kW heater running for 1 hour will use 1 kWh (the units recorded by a household electricity meter).

KEY POINT

In other units, you will see a U used as the symbol for voltage. This is because U is used in BS 7671 Requirements for Electrical Installations (the IET Wiring Regulations) and other European standards as the voltage between two parts of an electrical system.

ACTIVITY

Look in Part 2 of BS 7671 in the section called 'Symbols Used in the Regulations'. What is represented by:

a) U

b) U_0

c) U_{oc}.

Assessment criteria

2.3 Identify appropriate electrical instruments for the measurement and calculation of different electrical values

KEY POINT

An ammeter has a very low internal resistance and must never be connected across the supply.

ASSESSMENT GUIDANCE

Instruments may be either analogue or digital. Analogue instruments normally have a needle or pointer which follows the input value (analogous). A digital meter has a numeric display.

KEY POINT

As voltmeters have very high resistance they can be connected across the supply.

ACTIVITY

Why will a voltmeter not affect the circuit parameters (conditions) if connected across a resistor of low value, but will affect the readings if the resistor is of a high value?

MEASURING ELECTRICAL QUANTITIES IN CIRCUITS

In order to measure the correct values, ammeters and voltmeters need to be connected differently.

As ammeters measure the current flowing through a circuit, they are connected in series with the load.

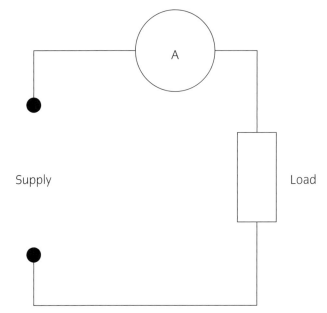

An ammeter (A) must be connected in series

As voltmeters read the potential difference across the load, they are connected in parallel.

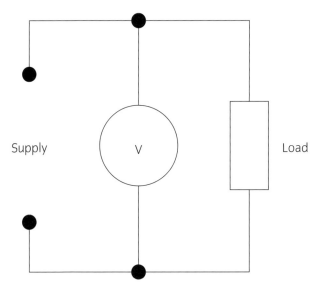

Volmeter (V) connection in parallel

Wattmeters measure the power in a circuit, which is determined by the current and voltage. As a result, they are connected in parallel and series, allowing measurement of both properties.

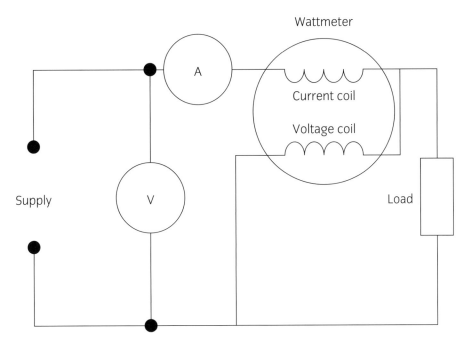

Wattmeter connection

Unlike the meters above, ohmmeters must only be used when a circuit is **not** energised. Ohmmeters use a small internal current and voltage to determine resistance and must be connected in parallel to the given resistance, as shown below.

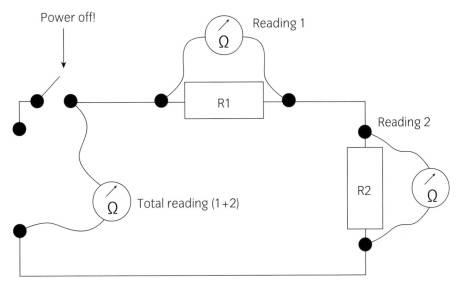

Three different ohmmeter readings with series resistors

SmartScreen Unit 309
PowerPoint 3 and Handout 3

Assessment criteria

3.1 Specify what is meant by mass and weight

Understanding basic mechanics will assist you to link the mechanical properties and electrical quantities that enable electricity to be turned into a useful form of energy, linking physical work and electrical power. It may also assist you to understand everyday objects, from seesaws to cranes and from claw hammers to wire cutters.

MEASURING MECHANICAL LOADS

It is important to understand the ways in which mechanical loads are measured. You need to understand the differences between mass and weight as well as energy and power. Energy and power can be mechanical as well as electrical.

The fundamental relationship between the mass and the weight is defined by Newton's Second Law and can be expressed as:

$$F = ma$$

where:

F = force (N)

m = mass (kg)

a = acceleration (m/s^2).

Weight

Weight is the gravitational force acting on a body mass. Newton's Second Law can be related to weight as a force due to gravity and can be expressed as:

$$W = mg$$

where:

W = **weight** *(N)*

m = **mass** (kg)

g = acceleration of gravity (on Earth this is 9.81 m/s^2).

Mass

Mass is the comparison of an amount of material measured against a known value. The SI unit of mass is the kilogram. At the International Bureau of Weights and Measures, located in Sevres France, a 1 kg mass of platinum is retained as the criterion for determining mass.

As mass is determined by balancing one body against another, gravitational pull does not make any difference to the value. Mass is the same in outer space as it is on Earth.

ASSESSMENT GUIDANCE

There is no weight in space, just mass.

Acceleration

Acceleration is the rate of change of velocity with time.

$$a = \frac{dv}{dt}$$

where:

a = acceleration
v = velocity
t = time.

As acceleration is a measurement of the rate of speed in metres per second (m/s) for every second, the unit is metres per second per second or metres per second squared (m/s^2).

CALCULATING MECHANICAL ADVANTAGE OF USING LEVERS

Assessment criteria

3.2 Explain the principles of basic mechanics as they apply to levers, gears and pulleys

Archimedes claimed that, if he was able to stand in the appropriate place, he could use a lever to move the Earth. Although it is not actually possible to lift the Earth, the principle is a good illustration of the power of levers.

How levers work

The total turning moment (**torque**) applied to the ends of a lever must be equal but opposite. Therefore:

force × distance = force × distance

Torque is measured in newton-metres (Nm). This is the product of the force (N) applied to a lever and the distance (m), from the **fulcrum**, of the point where the force is applied.

Torque

A force that causes rotation.

Fulcrum

The pivot point.

Levers are divided into three classes, as described below.

Class 1 levers

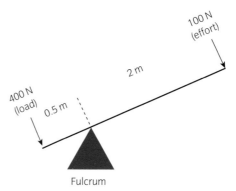

Example of a class 1 lever

In a class 1 lever, the force and the load are on different sides of the fulcrum. The effectiveness of the force is affected by the distance of the point of application of the force from the fulcrum. You measure the torque by multiplying the force by its distance from the fulcrum. So, the greater the distance, the greater the effect of the force. In the example above, force applied downwards can be four times less than the load as the lever on the force (effort) side of the fulcrum is four times longer than the lever on the load side. Examples of class 1 levers are seesaws, crowbars and claw hammers when used to lift nails.

The advantage gained by a lever is referred to as mechanical advantage (MA), which is calculated as:

$$MA = \frac{load}{effort} = \frac{400}{100} = 4$$

Class 2 levers

In a class 2 lever, the load is between the point of effort and the fulcrum. The calculations remain the same, but the load is more limited because it is between the two points, and the effort and the desired movement are both in the same direction. An example of a class 2 lever is a wheelbarrow.

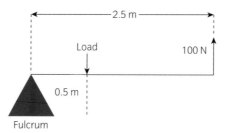

Example of a class 2 lever

In the example in the diagram above, the force of 100 N is applied at a distance of 2.5 m from the fulcrum and the load is applied 0.5 m from the fulcrum.

The load that can be lifted is calculated as follows:

force of load × distance of load from fulcrum
= force of effort × distance of effort from fulcrum

then:

$$0.5 \text{ m} \times \text{load} = 2.5 \text{ m} \times 100 \text{ N}$$

so:

$$\text{load} = \frac{2.5 \times 100}{0.5} = 500 \text{ N}$$

To find the mechanical advantage:

$$MA = \frac{\text{load}}{\text{effort}} = \frac{500}{100} = 5$$

Therefore the mechanical advantage is 5.

Class 3 levers

Class 3 levers are slightly different to class 2 levers in that the load is at the end opposite the fulcrum. Examples of class 3 levers include fishing rods, tweezers and the human arm.

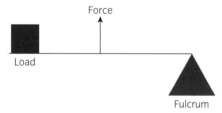

Force

Load

Fulcrum

Example of a class 3 lever

Pulleys

A pulley is an effective way of gaining a mechanical advantage when lifting an object. By running a rope through a four-pulley system, a mechanical advantage of four is gained.

Example

Determine the force required to raise the mass of 1000 kg.

As load has been described as a mass, first determine the downward force of the load. So:

force = mass × gravity

$F = 1000 \times 9.81 = 9810 \text{ N}$

(Remember: acceleration due to gravity on Earth is 9.81 m/s^2)

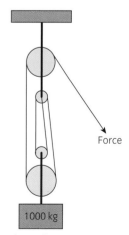

Force

1000 kg

Example of a pulley system

As the pulley system has four ropes compared to the one pulling rope, the mechanical advantage is 4:1. So:

$$\frac{9810}{4} = 2452 \text{ N}$$

So a downward force of 2452 newtons is required to raise the load.

Although the pulley gives a mechanical advantage, the pulling rope will need to be pulled four times further than the load is raised. This means, to raise the load by 1 m, the pulling rope needs to be pulled 4 m.

Gears

Gears are used to transmit torque from one rotating part to another. Gears are made of two or more cogwheels often having different diameters and therefore different numbers of teeth. Gears are used to change speed or effort needed to produce torque. For example, if a cog containing 20 teeth was connected to a cog containing 100 teeth and the smaller cog was driven by an electric motor rotating 40 times a second, we can determine the rotations per second of the large cog by expressing the gear ratio. So:

$$\frac{100}{20} = 5 \text{ giving a 5:1 ratio}$$

So if the small gear rotates five times more than the larger one, the number of times the larger wheel will rotate is:

$$\frac{40}{5} = 8 \text{ times per second}$$

CALCULATING QUANTITIES OF MECHANICAL LOADS

Force

Force can be defined as the strength or energy of physical action or movement. In practice, it can be any influence that causes an object with a mass to undergo a change in velocity, including any movement from rest, change in direction or geometrical construction (deformation).

Simply, force is also the 'weight' an object has due to the force of gravity pulling it onto the ground beneath it.

The SI unit for force, represented by the symbol F, is the newton (N). Since mass is measured in kilograms (kg) and acceleration is measured in metres per second per second (m/s^2), then:

$$\text{force in newtons} = \text{mass} \times \text{acceleration, so } F = ma \text{ (N)}$$

The force exerted on a mass by gravity can be determined for a mass of 1 kg, as acceleration caused by **standard gravity** on Earth is 9.81 m/s^2, so:

$$1 \text{ kg} \times 9.81 \text{ m/s}^2 = 9.81 \text{ newtons}$$

This can also be described as the weight of an object.

Example

To calculate how much force is needed to overcome gravity and lift a 500 kg object:

$$F = 500 \times 9.81 = 4905 \text{ N}$$

Example

Ignoring the downward force of gravity, if the above mass of 500 kg is accelerated horizontally at 1 m/s^2, the force working horizontally on the force is:

$$F = 500 \times 1 = 500 \text{ N}$$

Standard gravity

Standard gravity is the average gravity on Earth but the actual value can alter depending on where you are.

KEY POINT

On Earth, acceleration due to gravity is 9.81 m/s^2.

Work

If force is applied to a body and that results in movement, then work has been done. This applies to forces that lift or push objects or twist them.

When a force moves an object in the same direction as the force exerted, the work done is equal to the distance moved multiplied by the force exerted:

$$\text{work} = \text{force} \times \text{distance}$$

Or, to include the values used to determine force:

$$\text{work} = \text{mass} \times \text{gravity} \times \text{distance}$$

Mechanical work is measured in joules (J). (Newton-metres (Nm) can be used for mechanical work, but are also used as a measurement for **torque**.) Other units of work or energy, which are not SI units but commonly used for specific applications include:

- kilowatt hour (kWh), used to measure electrical energy by electricity supply companies
- calorie, often used as a measure of food energy
- BTU (British Thermal Unit), which is often used for heat source applications such as burning gas.

ASSESSMENT GUIDANCE

Work is done when a force applied to an object causes it to move. Lifting a glass full of liquid is therefore doing work.

Torque

A force that causes rotation.

Example

If a mass of 100 kg is lifted 10 m, calculate the work done.

Force = weight = mass × gravity = 100 × 9.81 = 981 newtons
= 981 N

Work done = force × distance = 981 × 10 = 9810 joules
= 9810 J

If the same mass is doubled, the work done is:

Force = weight = mass × gravity = 200 × 9.81 = 1962 N or 1.962 kN

Work done = force × distance = 1962 × 10 = 19620 joules

Energy can exist in many forms but is categorised in two main groups:

- kinetic energy, which is a force due to motion, such as a rotating machine
- potential energy, which is a force such as gravity or a spring.

For example, the potential energy of gravity keeps a mass on the ground. If the mass is to be raised, then a machine uses kinetic energy. If the input of kinetic energy ceases, the potential energy tries to bring the mass back down to the ground.

Power

Power is defined as the rate of doing work, ie work done divided by the time taken to carry out that work. The unit of power is joules per second (J/s), which is equivalent to a watt.

$$\text{average power} = \frac{\text{work done}}{\text{time taken}} \text{ (measured in joules per second or watts)}$$

Example

The output (mechanical) power required for a motor to raise a mass of 1000 kg to a height of 5 m above the ground in 1 minute is calculated as:

$$\text{power} = \frac{\text{mass} \times \text{gravity} \times \text{distance}}{\text{time}}$$

So:

$$\text{power} = \frac{1000 \times 9.81 \times 5}{60} = 817.5 \text{ watts}$$

If the same motor raised the same load in 10 seconds, the output power required by the motor would be:

$$\text{power} = \frac{\text{mass} \times \text{gravity} \times \text{distance}}{\text{time}}$$

So:

$$\text{power} = \frac{1000 \times 9.81 \times 5}{10} = 4905 \text{ watts}$$

Although the same amount of energy is used, no matter how quickly the task is carried out, the power required to do the work more quickly is increased due to the shorter time.

CALCULATING THE EFFICIENCY OF MACHINES

The law of conservation of energy states that energy cannot be created or destroyed. However, when energy is changed from one format to another, the energy does not fully transfer from one type to another. The remaining energy is converted into other forms, such as noise and heat, which are common causes of loss during energy transfer. Such losses, or wasted energy, are common in any mechanical process.

The efficiency of a mechanical system can be defined as the ratio of output power compared to the input power and is rated as a percentage. This can be expressed as:

$$\% \text{ efficiency} = \frac{\text{output energy or power}}{\text{input energy or power}} \times 100\%$$

It is more common for efficiency to be expressed in terms of power than energy.

Example

If a 200 kW output machine has an input power of 220 kW, the machine efficiency is:

$$\% \text{ efficiency} = \frac{200}{220} \times 100\% = 90.9\%$$

The passage of an electric current represents a flow of power or energy. When current flows in a circuit, power loss occurs in the conductors due to the conductor resistance. This manifests itself as heat dissipation and voltage drop.

POWER IN BASIC ELECTRICAL CIRCUITS

Like mechanical power, electrical power is also measured in watts and is also a measure of energy used over a period of time. Before you explore electrical power, you must be able to understand basic electrical principles. Therefore electrical power will be covered in greater detail in Learning outcomes 4 and 7.

ACTIVITY

A 4 kW electric motor drives a hoist. If the motor is 75% efficient and the hoist is 55% efficient, what is the output power of the hoist?

Assessment criteria

3.4 Calculate values of electrical energy, power and efficiency

Understand the relationship between resistance, resistivity, voltage, current and power

SmartScreen Unit 309
PowerPoint 4 and Handout 4

Assessment criteria

4.1 Describe the basic principles of electron theory

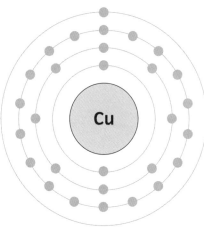

A simplified copper atom structure with electrons

KEY POINT

Atoms are bound by an electrical force, whereas molecules are bound by a chemical force.

ACTIVITY

Copper has 29 electrons and 29 protons. Using textbooks or the internet, find how many there are for:

a) carbon

b) aluminium

c) silicon

d) gold.

The basic principles of electricity have been studied for centuries and what is now common electrical theory was once groundbreaking new discovery. You need to understand the basic principles, including atomic composition, in order to work safely with electricity, magnetism and electrochemical reactions, and to progress in the industry.

ATOMIC THEORY

In order to understand where electricity comes from and what it is, it is necessary to understand a small amount of atomic theory.

Atoms are very small particles that are sometimes arranged as molecules. An atom is not solid but is made up of smaller particles, separated by space. The centre of an atom is the nucleus, which is made up from various particles including protons and neutrons. Protons are positively charged and neutrons have no charge. Orbiting the atom are the electrons, which are negatively charged.

The atoms that make up different materials have different numbers of electrons. In the steady state an atom has equal numbers of protons and electrons, and this leaves the atom electrically neutral.

Atoms in solids and liquids are more tightly bound together than those in gases. The diagram (left) shows the simplified structure of copper, which is often used to conduct electricity. The representation is two-dimensional when in fact the actual atom is three-dimensional. Where there are two or more electrons orbiting a nucleus, their orbiting paths are known as shells. The electron paths (shells) form an elliptical orbit.

Reaction of atoms

Different atoms have different numbers of electrons. Copper has 29 electrons and 29 protons. The outer shell is weakly held in orbit and can break free, causing random movement among other copper atoms. The loss of an electron causes an atom to become positively charged. It is known as a positive ion. Positive ions attract electrons, causing electron movement. Negative ions have more electrons orbiting them than protons in the nucleus.

The movement of electrons throughout a material is random but, by the laws of electric charge, like charges repel and unlike charges attract.

The diagram below shows the random movement of electrons in a material. The inset shows the electrons orbiting the proton. Electrons on an outer shell are released, as the force of attraction by the proton is weak, and the electron moves to the next proton.

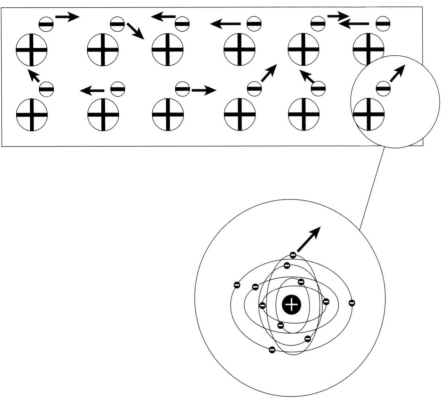

Random movement of electrons in a material

Flow of electrons

Random free electrons can be configured if the conducting material is connected to a battery. The free electrons are attracted to the positive plate and repelled by the negative plate. This causes the electrons to drift, in a conducting material, from the negative terminal of the battery to the positive terminal. As positive ions are unable to drift in solids, every time an electron leaves the negative terminal, one enters the positive terminal. This flow of electrons is electric current.

KEY POINT

Electron is the Greek word for amber. Amber is a material that is easily electrified by static. Rubbing wool on amber can charge the amber, which can then release an electric charge when held against another material or person.

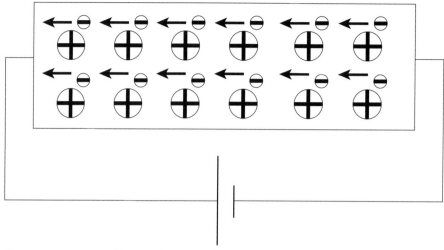

Electrons are attracted in one direction when a source of energy is connected to the material.

Closed circuit

A complete circuit connected to a source of energy. If the circuit contains a switch and the switch is switched off, it becomes an open circuit.

> **KEY POINT**
>
> Conventional current direction is actually opposite to actual current direction. Current flows from negative to positive but, in our industry, we still refer to conventional current flow.

> **KEY POINT**
>
> The flow of current in one direction is called direct current (d.c.), which is the main current form referred to in this unit.

In order for electric current to flow, there must be a **closed circuit**. Once the circuit is opened, the drift of electrons is immediately stopped and the current flow ceases.

Current direction

Electric current flows from a negative terminal to a positive terminal. However, before atoms and electrons were understood, scientists believed that electricity was a fluid and flowed from positive to negative. This is known as conventional current direction and, although it is now understood that electrons flow from negative to positive, we still refer to conventional current direction as being positive to negative.

How many electrons make one ampere?

The flow of electrons in one direction is known as charge, which is measured in coulombs (C).

As electron flow is electrical current and current is measured in amperes (A), how is one converted to the other?

French physicist Charles Coulomb (1736–1806) determined that 1 coulomb of charge is equal to 6.24×10^{18} electrons. That is equal to 6 240 000 000 000 000 000 electrons! So if that many electrons flowed through a material such as copper and past an electron counter, that would be equal to 1 coulomb. He also determined that if the drift of electrons was at a rate of 1 coulomb per second, the resulting current would be 1 ampere.

Therefore a current of 1 ampere flowing indicates a charge of 1 coulomb per second, giving:

$$Q = It \text{ or } I = \frac{Q}{t}$$

where:

Q = charge transferred, in coulombs (C)
I = current, in amperes (A)
t = time, in seconds (s).

Example

If a total charge to be transferred is 750 C in one minute, the current flow is calculated as follows.

Using the formula $Q = It$, calculate the current:

$$I = \frac{Q}{t}$$

Since 1 minute = 60 seconds (s): $\dfrac{750}{60} = 12.5$ A

The current flow is therefore 12.5 A for 60 seconds to give the total charge of 750 C.

If a current of 25 A was to flow for 2 minutes in the above circuit arrangement, the total charge would be calculated like this.

Since 2 minutes = 120 s and $Q = It$:

$$Q = It = 25 \times 120 = 3000 \text{ C}$$

Therefore the total charge would be 3000 coulombs.

INSULATORS AND CONDUCTORS

As has been described in Assessment criteria 4.1, the movement of free electrons constitutes the flow of electric current and since the atomic structure varies from material to material, some will allow electron movement better than others when an external voltage is applied. Where the freedom of electrons to move is high, the material will act as a good conductor of electricity.

Examples of conductors are:

- aluminium and copper (used in cables and overhead line conductors)
- brass (used in plug pins and terminals)
- carbon (motor brushes)
- mercury (discharge lamps and special contacts)
- sodium (discharge lamps)
- tungsten (lamp filaments).

ASSESSMENT GUIDANCE

These equations may look strange, but do not let them put you off. The letters stand for numbers. Simple calculations are all you will need to do.

KEY POINT

Where a formula shows two symbols together with no mathematical symbol, it means they must be multiplied. So $Q = It$ simply means $Q = I \times t$.

ACTIVITY

Remember that the SI unit of time is the second, not the minute, hour or day. Use the internet to find the definition of the second.

Assessment criteria

4.2 Identify and differentiate between materials which are good conductors and insulators

However, where the material's atomic structure is such that there is minimal electron movement, there will be negligible current flow and the material will act as an insulator.

Examples of insulators are:

■ thermoplastics and polyethylene (cable insulation)

■ glass and porcelain (overhead line conductor support insulators)

■ rubber (mats, gloves and shrouding for live working).

It must be stressed, however, that the level of insulation afforded by an insulator can be severely reduced by:

■ damage (cracks, splits, etc)

■ deterioration (cracks, splits, etc due to ageing)

■ contamination (water, salt spray, chemicals, etc).

CABLE COMPONENTS AND TYPES

Electric cables used for electrical installations in industrial, commercial and domestic situations come in a wide range of sizes, materials and types. Electric cables used for the long-distance transmission of electrical energy (400 kV and 275 kV) are normally buried in the ground or suspended on towers or pylons and the cables used for the more local distribution of electrical energy around the country (132 kV, 66 kV, 33 kV and 11 kV) are buried in the ground or suspended on towers or pylons or on wooden poles.

Cables generally consist of three major components:

■ the current-carrying material (conductors)

■ electrical insulation (normally colour or number coded for identification)

■ a protective outer covering called a sheath (this is not present on some single-core cables).

The makeup of individual cables varies according to the application for which they are to be used. The construction and material are determined by three main factors:

■ working voltage, which determines the thickness of the insulation

■ current-carrying capacity, which determines the cross-sectional area of the conductor(s)

■ environmental conditions – such as requirement for mechanical protection, temperature, water and chemical protection – which ultimately determine the form and composition of the outer protective sheath.

The current-carrying conductors of an electricity cable are normally made of copper or sometimes aluminium, either of stranded or solid construction.

Assessment criteria

4.3 State the types and properties of different electrical cables

ASSESSMENT GUIDANCE

The armouring of cables is there to provide mechanical protection to the inner conductors. It is normally made of steel wire in a single or double layer wrapped around the cable bedding and overlaid with PVC sheathing. In older cables steel tape may be found.

SmartScreen Unit 309

Handout 13

Cable types

There are generally two types of cable:

- single-core cables
- multi-core composite cables.

Single-core cables

Single-core cables are generally un-sheathed cables with only one core or conductor. Insulated single-core cables are required to be installed in containment systems such as conduit or trunking in order to provide mechanical protection. The insulation will usually be colour coded to identify the conductor. Single-core cables may also be bare, for example in the case of supply cables, but these must be suspended from pylons. Single-core cables may also have a further sheath, these are commonly called double-insulated cables.

PVC/PVC cables

In industrial, commercial and domestic wiring installations PVC-insulated and PVC-sheathed cable is the most common form of cable used and is either surface mounted by clipping, embedded in the fabric of the building (plaster), or installed in conduit or trunking.

The conductors are covered with colour-coded PVC insulation and then enclosed in a PVC outer sheath.

PVC/PVC cable (also known as twin and earth)

PVC/SWA cable

Cables for direct burial or for exposed installations are constructed of stranded conductors covered with colour-coded PVC insulation and then enclosed in an inner PVC sheath. Steel wire armour (SWA) in the form of steel wires spiralled around the cable provides mechanical protection and an outer sheath of PVC provides corrosion protection. The SWA can provide the circuit protective conductor (CPC) and the cable is terminated by the use of a compression gland. If the SWA does provide the CPC, all strands of the armour must be terminated in the gland and the gland effectively earthed.

SWA with gland assembly

Flexible cable (Flex)

Flexible cable (flexes) connect electrical appliances to the mains, either via plugs or switched spurs. There are many sizes and types of flexible cable but essentially they are all made up of two or three separately insulated cores and in each of these cores the conductor is made up of many thin strands of copper conductors, which give the cord its flexibility.

In three-core flex the cores are colour coded in accordance with European Harmonisation document HD 308: Identification of cores in cables and flexible cords, and to align with BS EN 60445: 2010 Basic and safety principles for man-machine interface, marking and identification – identification of equipment terminals, conductor terminations and conductors. The colours are: brown for live, blue for neutral, green and yellow striped for earth. Three-core flex must be used for equipment that relies upon an earth for safety (class 1).

Two-core flex has only the live and neutral conductors, coloured brown and blue. The outer sheath of a flex can be PVC or rubber or rubber/textile braided. Two-core flex can have a round or flat sheath.

MI cable

Mineral insulated cable (MI cable) is made by placing copper conductors inside a seamless copper tube filled with dry magnesium oxide powder. The cable has a small overall diameter and, apart from the terminations, does not use organic material as insulation. This makes the cable more resistant to fires than PVC-insulated cables and, as such, they are used in hazardous areas, such as oil refineries and, chemical works, and critical fire protection applications, such as alarm circuits, fire pumps and smoke control systems. MI cable is often known as MICC (mineral insulted copper clad) and pyro after the original manufacturer Pyrotenax.

MI cables may be covered with a plastic sheath, coloured for identification purposes:

- orange for general electrical wiring
- red for fire alarm wiring
- white for emergency lighting.

The plastic sheath also provides additional corrosion protection for the copper sheath.

The copper sheath provides the CPC and the cable is terminated, using a sealing pot filled with sealing compound. The pot is fixed to the sheath, using a self-tapping thread. Terminating MI cable requires quite a high skill level to ensure the termination is electrically sound.

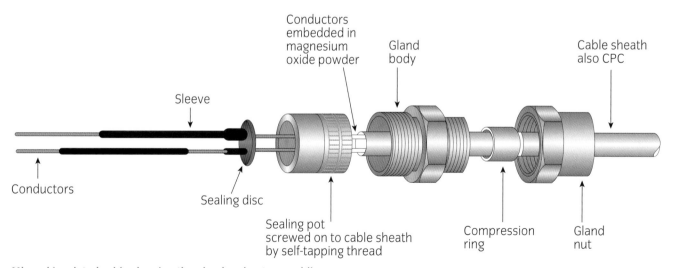

Conductors embedded in magnesium oxide powder

Gland body

Cable sheath also CPC

Sleeve

Conductors

Sealing disc

Sealing pot screwed on to cable sheath by self-tapping thread

Compression ring

Gland nut

Mineral insulated cable showing the gland and pot assemblies

Fire performance cable

Fire performance cables maintain circuit integrity in a fire and so are suitable for fire detection, alarm and emergency lighting systems. The cable is usually manufactured with solid or stranded conductors that have a fire barrier of a proprietary fire-resistant material such as 'insudite' or mica glass tape. This alternative to MI cable does not require special terminations and, as such, is much easier to install than MI cable.

Optical fibre cable

Optical fibre cables are communication cables containing one or more optical fibres and are used for telecommunications and computer networking. The optical fibre elements, made from optical quality plastic, are individually coated with plastic layers and contained in a protective tube. The energy is passed down the cable in the form of a digital pulse of laser light, which always stays within the optical fibre. Each fibre can carry a great number of independent channels, each using a different wavelength of light.

High voltage power cables

The IEC (International Electrotechnical Commission) define a.c. high voltage as above 1000 V. Electric cables used for the long-distance transmission of electrical energy (400 kV and 275 kV) are normally buried in the ground or suspended on towers or pylons and the cables used for the more local distribution of electrical energy around the country (132 kV, 66 kV, 33 kV and 11 kV) are buried in the ground or suspended on towers or pylons or on wooden poles.

Generally, electricity cables are laid directly in the ground and at sufficient depth to avoid undue interference or damage. High voltage power cables operating at 11 kV, which is the common voltage used for underground electricity distribution in urban areas, are required to be placed at a depth of 600 mm. Most power cable sheaths are coloured black, though some high voltage cables are red. If ducts are used and of modern plastic construction, they are normally coloured black.

It was shown in earlier learning outcomes that a given amount of power can be transmitted along a conductor either at a high current and low voltage or at a low current and a high voltage. For economic reasons, the latter is the preferred approach since transmission at low currents means that smaller cross-sectional area conductors can be used as well as the power loss being significantly less.

However, the use of high voltages requires a high level of electrical insulation for the live conductors. High voltage transmission is considerably cheaper by overhead lines rather than underground cables since it is easier to provide the necessary high level of insulation required. For this reason the majority of the National Grid uses overhead transmission (at voltages up to 400 kV).

ACTIVITY

Using the internet, read about the early battles between Edison and Westinghouse over which supply would win (d.c. or a.c.). It is reckoned that Edison would have needed a power station on each street corner.

SmartScreen Unit 309
Handout 26

Assessment criteria

4.4 Describe what is meant by resistance and resistivity in relation to electrical circuits

ACTIVITY

Have you ever wondered why cable sizes are as they are? Why don't they go up in even steps?

SmartScreen Unit 309

Handout 14

RESISTANCE AND RESISTIVITY

As we have seen, the flow of electric current in a material is related to its atomic structure and the freedom of electrons to move. Since the atomic structure varies from material to material, some will allow electron movement better than others when an external voltage is applied. Where the freedom of electrons to move is high, the material will act as a good conductor of electricity. However, where the material atomic structure is such that there is minimal electron movement, there will be negligible current flow and the material will act as an insulator.

Resistance in an electrical circuit is encountered in two forms: continuity and insulation resistance, corresponding to the two functions mentioned above, ie conductor and insulator.

Resistance of a conductor

This is the end-to-end resistance of an electrical conductor or the resistance between two points in an electrical circuit. In the case of an electrical conductor, the continuity resistance needs to be as low as possible to enable the maximum amount of current to flow through the conductor. Continuity resistance is generally measured on an ohmmeter. Measured values can be very small and therefore may be quoted in milliohms (mΩ).

The resistance of a conductor is based on a number of variable factors:

- type of conductor
- physical dimensions
- temperature.

In order to understand the resistance of a conductor, it is necessary to be aware of these factors.

During the design and any inspection and testing of an electrical installation it is always important to keep these factors in mind.

Designers will factor in temperature increases for conductors because the temperature of a conductor under load will increase, resulting in increased conductor resistance.

Inspectors, whilst testing circuit resistances under no-load conditions, need to understand that the actual resistance will be greater when the circuit is carrying a full load current. They will need to factor in the temperature increases.

Effects of physical dimensions: length

If a cube of conducting material, with a resistance between two opposite faces of R ohms, is joined in a line of equal cubes, then the overall resistance is that of all the cubes added together.

This arrangement is known as resistances in series or a series circuit, for example:

$R_{t_{series}} = R_1 + R_2 + R_3 + R_4 + R_5 = 5 \times R = 5R$, as shown in the diagram.

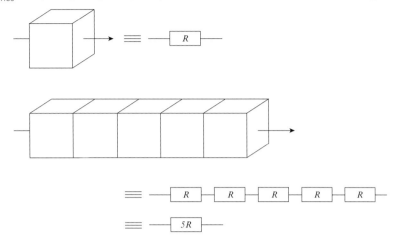

The effect of increasing length on resistance

ASSESSMENT GUIDANCE

Practise doing series and parallel combination calculations, finding the unknown in a group of resistors, as well as the total resistance.

KEY POINT

The more length that is added to a series circuit, the greater the resistance. Resistors in a series arrangement are added to each other.

KEY POINT

R_t is sometimes referred to as R_e. This would be the equivalent resistance if all the individual resistances were replaced by a single resistor.

Example

If there are seven pieces of conducting material (resistors) connected in a series arrangement, the formula would be:

$$R_{t_{series}} = R_1 + R_2 + R_3 + R_4 + R_5 + R_6 + R_7$$

If the value of each resistor is the same, the total is $7 \times R = 7R$.
If $R = 2\,\Omega$, $7R = 14\,\Omega$.

If the resistors have different values, they are still added together.
If $R_1 = 2\,\Omega$, $R_2 = 3\,\Omega$, $R_3 = 4\,\Omega$, $R_4 = 6\,\Omega$, $R_5 = 9\,\Omega$, $R_6 = 3\,\Omega$, $R_7 = 1\,\Omega$, then:

$$R_{t_{series}} = 2 + 3 + 4 + 6 + 9 + 3 + 1 = 28\,\Omega$$

As conductors in cables are generally the same diameter throughout their length, each metre of cable is like one of the cubes in the diagram above. The longer the conductor, the higher the resistance.

Example

If a conductor in a cable has a resistance of $0.01\,\Omega$ per metre (Ω/m), the resistance of 120 m is calculated as follows.

As the total resistance (R_t) = $\Omega/m \times$ length, then $0.01 \times 120 = 1.2\,\Omega$

Effects of physical dimensions: cross-sectional area

If a pipe had a cross-sectional area of 150 mm^2 and water was poured into it, the flow of water would be restricted by the cross-sectional area of the pipe as only so much water could pass through it at one time. If the pipe diameter was increased to 300 mm^2, twice the amount of water would be able to flow. Current flowing through a cable is much the same. The smaller the cross-sectional area of a conductor, the greater the resistance. If the cross-sectional area is increased, the resistance decreases proportionally. This effect is the same as having resistances in parallel, which is covered on page 43.

By increasing the cross-sectional area by a factor of 4, the resistance is reduced accordingly.

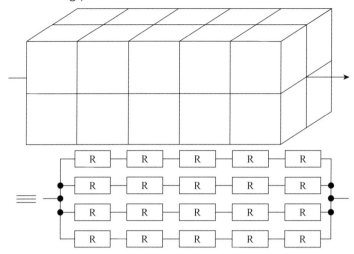

The effect of increasing the cross-sectional area

Example

If the resistance of 100 m of a particular cable is 0.3 Ω, the resistance for 1 km of the same cable, assuming the temperature remains constant, is calculated as follows.

1 km = 1000 m

Therefore, if the resistance of 100 m = 0.3 Ω:

the resistance of 1000 m = 10 × the resistance of 100 m = 10 × 0.3 = 3 Ω

Example

If the cross-sectional area of the cable is doubled in size, the resistance would halve, so:

$$\frac{3\,\Omega}{2} = 1.5\,\Omega$$

Cable manufacturers' data provides resistance values for cable conductors according to the conductor material and cross-sectional area, usually expressed in mΩ/m at a specified temperature. If the end-to-end cable length is known, the end-to-end resistance can then be calculated.

INSULATION RESISTANCE

This is the resistance measured across the electrical insulation surrounding a conductor, from the outside of the cable through to the conductor or through the insulation separating conductors. In this case the resistance needs to be very high to prevent current leakage from the circuit or a short circuit between adjacent circuit conductors. Insulation resistance is measured on an insulation resistance tester, giving readings in megohms (MΩ). Older instruments provided with a scale and pointer display usually have the scale marked with a range of 0–∞ MΩ. Since the measured value will invariably be very high, the pointer is likely to swing full scale to the ∞ (infinity) mark. This does not mean that the insulation resistance is infinity! The measurement is beyond the scale range of that particular instrument and, to be correct, the reading should be recorded as greater than the highest marked scale value or maximum of scale range if known, eg > 99 MΩ (greater than 99 MΩ).

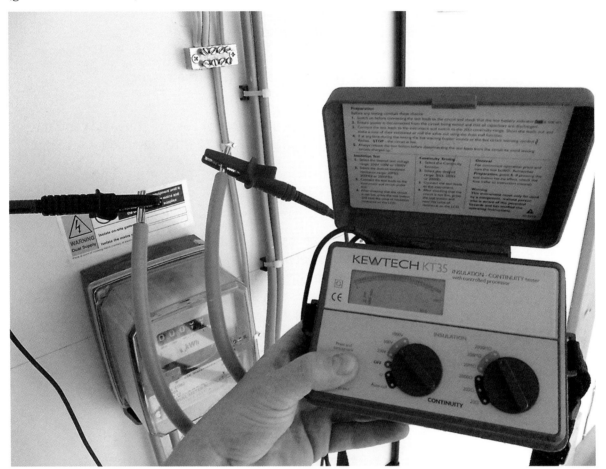

Testing between live conductors

A high reading (typically > 1 MΩ) indicates a good standard of electrical insulation around live parts. This will result in negligible current leakage and so the circuit is described as 'healthy'. A low reading (eg less than 0.5 MΩ) indicates deterioration in the effectiveness of the insulation, leading to the possibility of a fault in the form of a short circuit across the circuit conductors.

Effects of conductor materials: specific resistivity

The specific resistance of a material is normally very low and is expressed in microhms (μΩ). The resistivity of a material is expressed in the format μΩm. For example, the resistance of a one-metre cube of copper is 0.0172 μΩm.

KEY POINT

Values of resistivity for materials used as conductors are always based on resistance at 20°C.

ACTIVITY

Put these conductors in ascending order of resistivity: copper, steel, aluminium, brass, silver, gold.

SI units

The units of measurement adopted for international use by the *Système International d'Unités*.

The specific resistance of a material is represented by the Greek letter rho (ρ). Therefore the resistance of a particular conductor can be calculated as:

$$R = \frac{\rho l}{A}$$

where:

R = conductor resistance

ρ = cable resistivity in ohms per metre (Ω/m)

l = length in metres (m)

A = cable cross-sectional area in square metres (m^2).

Example

To calculate the resistance of 1000 m of 16 mm^2 annealed copper cable, where ρ = 0.0172 $\mu\Omega$m (or 0.0172 × 10^{-6} Ωm) and A = 16 mm^2 (or 16 × 10^{-6} m):

$$\text{use } R = \frac{\rho l}{A} = \frac{0.0172 \times 10^{-6} \times 1000}{16 \times 10^{-6}} = \frac{17.2 \times 10^{-6}}{16 \times 10^{-6}} = \frac{17.2}{16} = 1.075\,\Omega$$

Traditionally, in formulae the base SI units are always used (metres not millimetres, and ohms not microhms). Where the top line of a formula contains a 10^{-6} and the bottom line contains a 10^{-6} these values cancel each other out. This means that this formula is always an exception to the rule. As long as the value of resistivity used is microhms per metre and the cross-sectional area is in square millimetres, the two values can be input directly as the factors of 10^{-6} applied to both quantities always cancel out. So:

$$\frac{0.0172 \times 1000}{16} = 1.075\,\Omega$$

Effects of temperature

Generally, as the temperature of a material increases, so too the resistance of that material increases.

This variable is determined using the temperature coefficient of resistance, represented by the Greek letter alpha (α). For example, the coefficient for copper at 0 °C is 0.0043 Ω/°C and is represented by R_0. The coefficient for copper at 20 °C is 0.003 96 Ω/°C.

$$R_{t_1} = R_{t_2}(1 + \alpha T)$$

where:

R_t = resistance at the new temperature

R_0 = resistance at a given temperature

α = temperature coefficient

T = temperature change.

Example

If the resistance of 100 m of 2.5 mm^2 annealed copper cable is 0.5375 Ω at 20 °C and the coefficient α for copper at 20 °C is 0.003 96 Ω/°C, its resistance at 50 °C is calculated as:

$$R_{50} = R_{20}(1 + \alpha T) = 0.5375(1 + 0.003\,96(50 - 20))\,\Omega$$
$$= 0.5375 \times 1.1188\,\Omega$$
$$= 0.60\,\Omega$$

In electrical installations, it is important to be able to calculate conductor resistance at the standard room temperature of 20 °C as well as at the conductor's working temperature, which may be 70 °C. In many publications used by electricians, tables used to calculate the resistance of circuits are all based on resistivity and factors used to adjust temperatures are based on temperature coefficients.

Many of these tables give factors for correcting resistance to a given temperature. On closer inspection of these tables, it can be seen that the resistance of a copper conductor changes by 2% for every five-degree change in temperature.

If a conductor has a resistance of 0.5 Ω at 20 °C and the temperature is increased by 15 degrees, the resistance will increase by 6%. So the new resistance will be:

$$0.5\,\Omega \times 1.06 = 0.53\,\Omega$$

Multiplying the value of the resistance by 1.06 will increase it by 6%. To reduce the value of the resistance by 6%, it must be multiplied by 0.94.

Example

A 10 mm^2 conductor is 27 m in length and has a resistivity of 0.0172 μΩm at 20 °C.

Calculate the resistance of the conductor at 70 °C.

$$\text{Resistance at } 20\,°C = \frac{\rho l}{A} = \frac{0.0172 \times 27}{10} = 0.046\,\Omega$$

If the temperature increases to 70 °C, the change is 50 degrees. At 2% for every five degrees, this is a 20% increase, so apply a multiplier of 1.2.

$$0.046\,\Omega \times 1.2 = 0.056\,\Omega$$

APPLYING OHM'S LAW

One of the most used electrical formulae is Ohm's law. It states that a potential difference across a conductor is proportional to the current passed through it.

This proportionality is equal to the resistance (R) of the conductor. Therefore $V = I \times R$, which is generally written as $V = IR$.

This gives the potential difference (V), given the current and resistance.

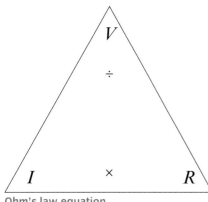

Ohm's law equation

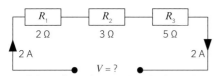

Applying Ohm's law to circuits in series

SmartScreen Unit 309
PowerPoint 15 and Handout 15

ACTIVITY

Ohm's law is normally quoted as the current being directly proportional to the applied voltage and inversely proportional to the resistance. The law is associated with Georg Simon Ohm 1789–1854. Whether he used this form of words is not known. Memorise Ohm's law as it forms the basis of many calculations you will use as an electrician.

To calculate the resistance, divide both sides of the equation by I to get:

$$R = \frac{V}{I}$$

You can remember the three variations of the Ohm's law formula by using a triangle. Covering the value that needs to be found leaves the correct calculation visible. So if resistance is needed when current and voltage are known, cover resistance and V over I (V divided by I) is left. Likewise, to determine V, cover it and you are left with $I \times R$.

Series circuits

Before considering Ohm's law to determine values of a series circuit, consider the characteristics of resistors in series. The total resistance of resistors in series is found by simply adding them together, so:

$$R_{total} = R_1 + R_2 + R_3 \ldots$$

As the current has to flow through all the resistors in the series circuit, the current is the same through all, ie it is constant.

Example

If a current of 2 A is passing through each of the resistors in a series circuit, the voltage at the terminals of the supply is calculated as follows.

$$\text{If } R_1 = 2\,\Omega, R_2 = 3\,\Omega \text{ and } R_3 = 5\,\Omega$$

then:

$$R_{total} = R_1 + R_2 + R_3 = 2 + 3 + 5 = 10\,\Omega$$

To calculate the voltage across the supply terminals:

$$V = IR = 2\,\text{A} \times 10\,\Omega = 20\,\text{V}$$

As the current in a series circuit is constant, to calculate the potential difference across each resistor, assuming no resistance in any of the connections and no internal impedance in the supply, use Ohm's law on each resistor, where V_1 is the potential difference across R_1 and so on.

$$V_1 = IR_1 = 2 \times 2 = 4\,\text{V}$$
$$V_2 = IR_2 = 2 \times 3 = 6\,\text{V}$$
$$V_3 = IR_3 = 2 \times 5 = 10\,\text{V}$$

This adds up to the 20 V across the terminals, showing that the effect of resistors in series is to form potential dividers.

Try the following examples of resistance in series. Remember these key points.

- The total resistance is equal to all the resistances added together.
- The current is constant in a series circuit.
- The voltages across each resistance, when added together, will equal the supply voltage for the circuit.

Example

Determine, for the circuit shown:

a) the total circuit resistance

b) the total circuit current (I_s).

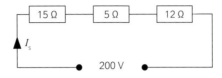

Example

Determine, for the circuit shown:

a) the total circuit resistance

b) the total circuit current (I_s)

c) the voltage drop across resistor 3 as indicated by the voltage meter V_3.

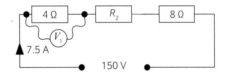

Example

Determine, for the circuit shown:

a) the total circuit resistance

b) the resistance of R_2

c) the voltage drop across resistor 1 as indicated by the voltage meter V_1.

Parallel circuits

With parallel circuits, the rules change. The total resistance is now found by using reciprocals, as shown.

$$\frac{1}{R_{total}} = \frac{1}{R_1} + \frac{1}{R_2} + \frac{1}{R_3} \ldots$$

In parallel circuits, the overall (total) resistance will be lower than the lowest resistance in the circuit. This is because the current can flow through the lowest resistance path as well as the other paths, effectively lowering the resistance.

Unlike series circuits, in parallel circuits voltage becomes the constant, with current varying across each resistor. The total circuit current is equal to the value of current flowing through each resistor, all added together.

ACTIVITY

In a parallel circuit, the greater the cross-sectional area the lower the resistance. The reciprocal of a resistance, $\frac{1}{R}$, is the conductance (ability to conduct) and conductances can be added together. When the total conductance is found, the reciprocal of this will be the total resistance. The unit of conductance is the siemens (S).

Memorise the formula for the total resistance of parallel circuits and use it in calculations.

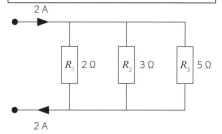

Applying Ohm's law to circuits in parallel

Common denominator

A denominator that can be divided exactly by all of the denominators in the question.

Example

If the same resistors ($R_1 = 2\,\Omega$, $R_2 = 3\,\Omega$ and $R_3 = 5\,\Omega$) are applied in parallel and the current flowing through the source is 2 A, calculate:

■ the voltage at the terminals

■ the voltage drop across each terminal of the resistors

■ the current passing through each resistor.

As the value of current through the source is known, calculate the voltage of the source by determining the total resistance value.

$$\frac{1}{R_{total}} = \frac{1}{R_1} + \frac{1}{R_2} + \frac{1}{R_3} \cdots$$

This can be approached in several ways. First, find a **common denominator**, in this case 30. Then work out how many times each of the original denominators divides into 30.

$$\frac{1}{R_{total}} = \frac{1}{2} + \frac{1}{3} + \frac{1}{5} = \frac{15 + 10 + 6}{30} = \frac{31}{30}$$

This gives the reciprocal of the total resistance, so turn both sides of the equations upside down.

$$R_{total} = \frac{30}{31} = 0.967\,\Omega$$

Alternatively, you can use the button marked on a calculator as x^{-1} and apply:

$2\,x^{-1} + 3\,x^{-1} + 5\,x^{-1} = x^{-1} =$ and the answer should be 0.967 Ω.

Always remember to push the final x^{-1} to get the true value.

To calculate the voltage across the supply terminals, use:

$$V = IR = 2 \times 0.967 = 1.934\text{ V (2 V)}$$

As the voltage in a parallel circuit is common, the voltage drop across each resistor is 2 V.

To calculate the current through each resistor, where I_1 is the current through resistor R$_1$ etc, apply the appropriate form of Ohm's law.

$$I_1 = \frac{V}{R_1} = \frac{1.934}{2} = 0.967\text{ A}$$

$$I_2 = \frac{V}{R_2} = \frac{1.934}{3} = 0.645\text{ A}$$

$$I_3 = \frac{V}{R_3} = \frac{1.934}{5} = 0.387\text{ A}$$

The total current flowing through the source is 2 A.

The passage of an electric current represents a flow of power or energy. When current flows in a circuit, power loss occurs in the conductors due to the conductor resistance. This manifests itself as heat dissipation and voltage drop.

POWER IN BASIC ELECTRICAL CIRCUITS

Assessment criteria

4.7 Calculate values of power in parallel and series d.c. circuits

It is known that if a potential difference of 1 volt exists between two points, then 1 joule of energy is used in moving 1 coulomb of electricity between the points. Therefore:

$$1 \text{ joule} = 1 \text{ coulomb} \times 1 \text{ volt or } W = QV$$

We also know that power is energy used over a period of time. In electrical circuits we also determine power using the amount of potential difference and current. So:

$$P = VI$$

SmartScreen Unit 309

PowerPoint 4 and Handout 4

Where:

P is electric power, in watts (W)

I is the current, in amperes (A)

V is the potential difference, in volts (V).

Also, as voltage can be determined using Ohm's law, by multiplying the voltage and resistance, then:

$$P = (IR)I = I^2R$$

Power loss or consumption is proportional to the square of the current flow, ie I^2. Thus, assuming a constant resistance in a circuit, if the current doubles (due to a corresponding increase in voltage) there will be a four-fold increase in power.

Also as:

$$I = \frac{V}{R} \text{ and } P = V \times I$$

Then:

$$P = \frac{V \times V}{R} = \frac{V^2}{R}$$

Note that these formulae are not complete for a.c. circuits as a power factor needs to be taken into account. This is covered in Learning outcome 7.

So, using the derived formula, the following calculations are possible.

Example

If a resistor of 10 kΩ is connected to a 100 V d.c. supply, the power dissipated in the resistor is calculated using:

$$P = \frac{V^2}{R} = \frac{100 \times 100}{10\,000}$$

$$= \frac{10\,000}{10\,000} = 1 \text{ W}$$

ACTIVITY

Calculate the resistance of a 3 kW electric fire connected to a 230 V supply.

Calculate the working (hot) resistance of a 60 W 230 V lamp, using:

$$P = \frac{V^2}{R}$$

Therefore:

$$R = \frac{V^2}{P} = \frac{230 \times 230}{60}$$

$$= \frac{52\,900}{60} = 881.67\ \Omega$$

Example

To determine the power dissipated in a resistor when a current of 100 A passes through and voltage 100 V is applied to the circuit:

$$P = VI = 100 \times 100 = 10\ kW$$

Calculate the value of the resistance in the same circuit:

$$P = \frac{V^2}{R}$$

Therefore:

$$R = \frac{V^2}{P} = \frac{100 \times 100}{10\,000} = \frac{10\,000}{10\,000} = 1\ \Omega$$

Assessment criteria

4.8 State what is meant by the term voltage drop in relation to electrical circuits

VOLTAGE DROP

The flow of electric current through a conductor results in a drop in electrical pressure (voltage), referred to as voltage drop or volt drop, due to the resistance to the current flow presented by the conductor resistance. This loss of voltage represents a loss of energy, which is reflected in the generation of heat.

The voltage drop is determined by the amount of current flowing in a conductor and the resistance of that conductor. It is calculated from the product of the current flowing, measured in amps (A), and the conductor resistance, measured in ohms (Ω), which is simple Ohm's law.

$$\begin{aligned} \text{Voltage drop} &= \text{Current} \times \text{Resistance} \\ V &= I \times R \end{aligned}$$

Example

If a circuit had a total resistance of 0.5 Ω and the load at the end of the circuit had a current demand of 15 A, the voltage drop in the circuit would be:

$$0.5\ \Omega \times 15\ A = 7.5\ V$$

So if the supply voltage at the origin of the circuit was 230 V, the voltage at the load would be:

$$230\ V - 7.5\ V = 222.5\ V$$

If voltage drop is excessive, there may not be enough voltage at the end of the circuit for the load to operate correctly. In order to reduce the amount of voltage drop, the cable resistance must be reduced. This can be done by increasing the cross-sectional area of the circuit conductor.

EFFECTS OF ELECTRIC CURRENT

The three main effects of electrical current (similar to sources of electricity) are:

- thermal (heating)
- chemical
- magnetism.

In this section we deal with the thermal and chemical effects. Magnetism is covered in detail in Learning outcome 5 (page 49).

Thermal (heating)

When current flows in a wire, apart from the flow of electrons, there is a thermal effect; the wire starts to heat up. The amount it heats up depends on factors such as the cross-sectional area of the wire, the amount of current flowing and the material that the wire is made of. The heating effect of electricity is used in electric fires and other heaters. Variations of this heat effect are used to make light from light-bulb (lamp) filaments, which give off large amounts of light as they glow white hot as a result of the current passing through the thin filament.

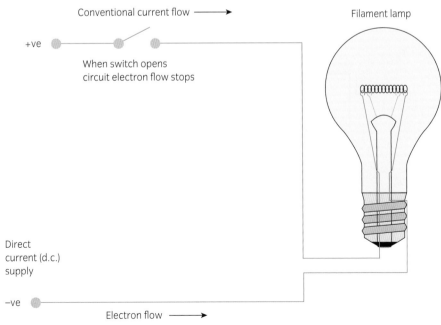

An electric light circuit using a d.c. source such as a battery

Assessment criteria

4.9 Describe the chemical and thermal effects of electrical currents

ACTIVITY

Early incandescent lamps used carbon filaments that were quite fragile. Unfortunately, carbon has a negative coefficient of resistance. This means that, as it gets hotter, the resistance goes down and so it will carry more current and become hotter still. This continues until the filament burns out. For this reason, carbon filaments were limited to small power ratings. What other material has a negative coefficient of resistance?

The effect of current passing through a wire and producing heat is a major consideration when designing electrical installations and will be covered at length during your course. Current that produces heat can be useful in electrical installations, for example in:

- electric heating
- lighting
- cooking
- circuit or equipment protection devices such as fuses or circuit breakers
- monitoring equipment.

There are also disadvantages, which include:

- circuit cables heating up, causing failure
- equipment getting too hot, causing danger
- energy loss.

A cable is designed to carry electricity from one place to another and is not supposed to heat up by any large amount. If it does heat up, it is using energy to do so which means less energy is available where it is required.

Chemical

When electric current is passed through a chemical solution, this causes basic chemical changes as ions are allowed to move to the positive plate, creating the process of electroplating or electrolysis. This process is used to coat material as, for example, in copper cladding on steel. If a copper-based solution were used, as shown in the diagram, the steel would become coated with copper.

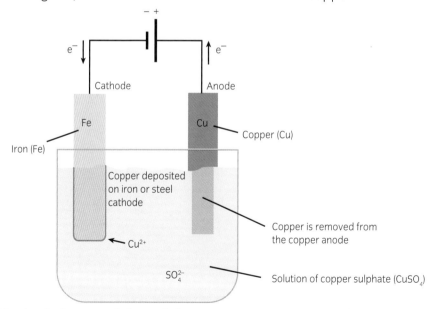

The chemical process of electroplating copper onto an iron or steel object

Understand the fundamental principles which underpin the relationship between magnetism and electricity

HOW ELECTRICAL COMPONENTS OPERATE

Electromagnets are often thought of as large magnets, found in vehicle salvage yards, lifting scrap metal from one place to another. Although this is one example, electromagnets are also used in lots of everyday items. For example, simply pressing a doorbell, or the release unit on a controlled access door, activates or de-activates an electromagnet. Without electromagnets many everyday tasks would be far more difficult.

First, consider what a magnet is.

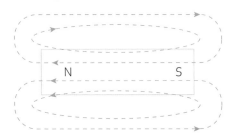

The pattern of magnetic flux lines that pass through a magnet from south to north

The bar magnet above shows a magnet and its north and south poles. It also shows the pattern of the lines of magnetic flux that pass through the magnet from south to north, and also outside the magnet from north to south. These flux patterns can be seen when a piece of paper is put over a bar magnet and iron filings are sprinkled over the paper. When the paper is tapped, the iron filings form a pattern because they are drawn into the flux lines.

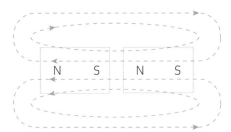

Two magnets, showing that opposite charges attract

When two magnets are put together, with a north pole facing a south pole, the lines of flux move together in the same direction. This causes

Assessment criteria

5.1 Describe the magnetic effects of electrical currents in terms of:

■ production of a magnetic field
■ force on a current-carrying conductor in a magnetic field
■ electromagnetism
■ electromotive force

SmartScreen Unit 309
PowerPoint 5 and Handout 5

ASSESSMENT GUIDANCE

Like poles repel, unlike poles attract.

the magnets to attract, pulling together and forming one larger magnet. Opposites attract.

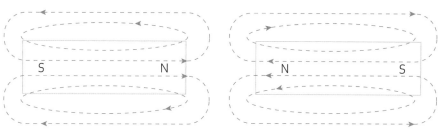

Two magnets, showing that like charges repel

When two magnets are placed with the same poles together, the flux paths move against each other. The force of the magnetic flux causes the magnets to repel and move away from each other.

The planet we live on is a giant magnet with a magnetic field. People navigate around the world, using this magnetic field by placing a small piece of iron on a pivot. Like the iron filings, this small piece of iron follows the flux direction. It is called a compass.

MAGNETIC FLUX PATTERNS OF ELECTROMAGNETS

Experimentation with a compass needle or iron filings on a sheet of paper with a conductor passing through it shows that a magnetic field is created around a conductor when current flows through it. If the current is removed, the effect on the compass or iron filings disappears.

This effect occurs throughout the length of a conductor. However, the effect on iron filings on a sheet of paper shows a 'slice' of the field in the plane where the paper is at right angles to the conductor.

Direction of current flow

Concentric rings of magnetic flux centre around the conductor

Current and field convention

It is usual to indicate current flow in a conductor because there is a three-dimensional relationship between current flow and magnetic field. Current flowing away from the viewer is shown with a cross, rather like an arrow or dart passing through a tube. Current flowing towards the viewer is shown as a large dot, like an arrow or dart point emerging from a hollow tube.

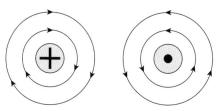

Direction of magnetic field around a conductor (shown in cross-section): current flowing away from view (left) and current flowing into view (right)

The direction of the magnetic field (field rotation) of the concentric rings can be checked with a compass needle. When current flows away from the viewer, the magnetic field rotates clockwise. When current flows towards the viewer, the magnetic field rotates anticlockwise. The magnetic field rotates in the same way as a screw: clockwise to tighten the screw (forcing it away), anti-clockwise to undo it (drawing it closer).

The strength of the magnetic flux is proportional to the current flowing through the conductor. The more current flowing, the stronger the magnetic field will be.

Placing two conductors together changes the effects. If two conductors are placed together in a conduit, for example, with the current flowing in opposite directions, there is a cancelling effect between the opposing magnetic fields, as long as the magnetic fields are of equal strength. This arrangement is therefore adopted in electrical installations. Magnetic fields can cause problems in electrical installations and therefore need to be cancelled and minimised as far as is reasonably practicable.

> **KEY POINT**
>
> Remember that parallel conductors with currents flowing in opposite directions will push away from each other. Currents flowing in the same direction will cause the conductors to pull towards each other.

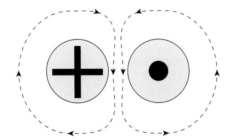

Cancellation effect of opposing conductors

Where conductors are placed together, with the current flowing in the same direction, there is an additional effect. This is undesirable in electrical installations as the increase in the magnetic field will cause additional losses in the circuit and possibly electromagnetic compatibility issues.

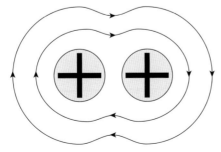

Totalling effect of magnetic fields

Solenoids

The strength of a magnetic field is proportional to the current flowing through the conductor. Even with high currents passing through the conductors, the field produced is relatively weak, in terms of useful magnetism. To obtain a stronger magnetic field a number of conductors can be added by turning or winding the cable.

The most common form of this is the solenoid, which consists of one long insulated conductor wound to form a coil. The winding of the coil causes the magnetic fields to merge into a stronger field similar to that of a permanent bar magnet. The strength of the field depends on the current and the number of turns.

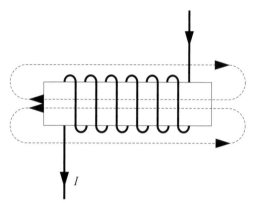

ASSESSMENT GUIDANCE

Winding the conductor on an iron core will considerably increase the strength of the magnetic field.

A cable wound around a tube: the current at the top moves away from the viewer and the current at the bottom moves towards the viewer

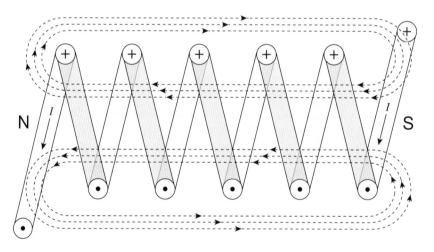

Coiling produces a bar magnet effect

As a solenoid is the electrically powered equivalent of a bar magnet, its field strength is dependent on the current passing through it. The magnetic field can be switched on or off.

The polarity of a solenoid is determined by the current direction. Using the NS rule, the letter N and/or S can be drawn, following the current direction.

The arrow heads on the letters, as shown in this diagram, indicate the direction of the magnetic field rotation.

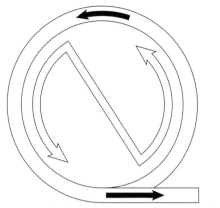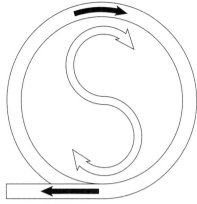

Tracing the letters shows the direction of the magnetic field rotation

The direction of magnetic field rotation can also be determined using the right-hand grip method. If the fingers of the right hand follow the current flow direction, the thumb points to the north pole.

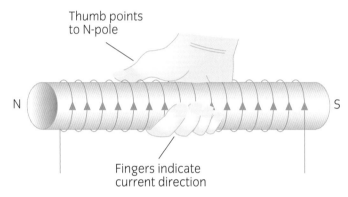

Thumb points
to N-pole

Fingers indicate
current direction

Holding a solenoid in a right-hand grip indicates the direction of magnetic field

Units of magnetic flux

The unit of magnetic flux is the weber (pronounced 'veyber'), abbreviated to Wb. It is represented by the Greek letter phi (Φ). Magnetic flux is a measure of the quantity of magnetic flux, not a density.

Flux density (the amount of flux in a given area) is represented by the symbol β, which is measured in webers per square metre (Wb/m^2), called teslas (T). One weber of flux spread evenly across a square metre of area gives a flux density of 1 tesla. Therefore:

$$\beta = \frac{\Phi}{A}$$

where:

β = magnetic flux density (T), in webers per square metre (Wb/m^2)

Φ = magnetic flux, in webers (Wb)

A = the cross-sectional area of flux path, in square metres (m^2).

ACTIVITY

The unit of magnetic flux is called the weber (Wb) and is named after Wilhelm Eduard Weber 1804–1891. The unit of magnetic flux density is called the tesla (T) and is named after Nikola Tesla 1856–1943. You should remember the units, what they represent and their symbols.

Force on a current-carrying conductor

If a current-carrying conductor is suspended in a magnetic field, the field induced by the current in the conductor reacts with the main magnetic field. This reaction of the two fields creates a force which moves the suspended conductor. The amount of force acting on the conductor is determined by:

$$F = \beta L I$$

where:

F = force acting on the conductor, in newtons (N)

β = magnetic flux density, in tesla (T), of the main field

L = the length of the conductor within the main magnetic field

I = the current passing through the conductor.

Example

A conductor 0.5 m in length is place in a magnetic field having a density of 0.5 tesla. If 10 A is passed through the cable, determine the force acting on the cable.

$$F = \beta L I$$

So:

$$F = 0.5 \times 0.5 \times 10 = 2.5 \text{ N}$$

ASSESSMENT GUIDANCE

The principle of force on a conductor underlies the operation of all motors. There are several hundred conductors in d.c. machines.

Assessment criteria

5.2 Describe the basic principles of generating an a.c. supply in terms of:

- a single-loop generator
- sine-wave
- frequency
- EMF
- magnetic flux

ACTIVITY

Think of some everyday things that contain alternators. One example is a wind-up torch, which uses an alternator to charge the batteries. Can you think of two others?

HOW ALTERNATORS PRODUCE SINUSOIDAL WAVEFORM OUTPUTS

Alternators use rotating magnets to generate electricity. When the magnetic field cuts through the conductor, which is wound on iron cores in the stator, an emf is produced and current is induced in the conductor.

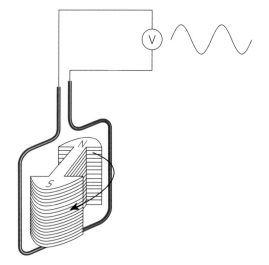

Simple alternator action

Calculating magnitudes of a generated emf

If you were to experiment with the electromagnetic induction apparatus shown on the previous page, you would find that when the magnet is changed for a stronger one, the flux increases, indicated by greater deflection on the meter. Furthermore, if the conductor can be doubled up by winding it into a coil, the deflection on the meter will be twice that of a single conductor. Speeding up the crossing of the magnetic field by the conductor also causes a greater deflection. The following formula can be deduced from these three effects:

$$E = \beta l v$$

where:

E = induced emf, in volts (V)

β = flux density of the magnetic field, in tesla (T)

l = the length of the conductor in the magnetic field, in metres (m)

v = velocity, in metres per second (m/s).

Example

If a conductor of length 0.1 m cuts a magnetic field of 1.2 T at a velocity of 15 m/s, the generated emf is calculated as follows:

$$E = \beta l v = 1.2 \times 0.1 \times 15 = 1.8 \text{ V}$$

If the conductor is twice the length (0.2 m), the emf is:

$$E = \beta l v = 1.2 \times 0.2 \times 15 = 3.6 \text{ V}$$

Static induction

Static induction is the induction of a current into a circuit where no physical movement has taken place.

The induction is caused by the rising or collapsing of the magnetic field effectively 'cutting' the conductor. The value of the statically induced emf depends on the change of total magnetic flux.

$$E = \frac{\Phi}{t}$$

and

$$\Phi = Et$$

where:

E = induced emf, in volts (V)

Φ = total magnetic flux change, in webers (Wb)

t = time for the flux change, in seconds (s).

Joule

The unit of measurement for energy (W), defined as the capacity to do work over a period of time.

Example

If a coil induces an emf of 250 V and it takes 10 ms for the current to fall to zero, the flux change is calculated as follows:

$$\Phi = Et = 250 \times 10 \times 10^{-3} = 2.5 \text{ Wb}$$

If the same coil takes twice as long (20 ms or 20×10^{-3} seconds), the change in the magnetic flux is:

$$\Phi = Et = 250 \times 20 \times 10^{-3} = 5 \text{ Wb}$$

SOURCES OF ELECTROMOTIVE FORCE

When electric current flows, energy is dissipated because it cannot be created or destroyed. As energy cannot be created, electrical energy has to be converted from an existing form of energy. The form of energy converted may be chemical, as in a battery, it may be mechanical, as in a generator, or a combination of materials reacting to a source of energy such as a solar photo-voltaic (PV) cell reacting to sunlight.

In the early days of electrical research, electricity was believed to be a fluid, which circulated as a result of an applied force. The term 'electromotive force' (emf) (E) was, and still is, used.

In determining units of electricity, emf is defined as the number of **joules** (J) of work required to move 1 C of electricity around a circuit. This unit of joules per coulomb is referred to as a volt (V).

$$1 \text{ volt (V)} = \frac{1 \text{ joule (J)}}{1 \text{ coulomb (C)}}$$

Example

If a battery of 12 V gives a current of 5 A for 10 minutes, the amount of energy provided over the 10-minute period is calculated as follows.

To find total energy: $W = Q \times V$

Total charge transferred:

$$Q = It = 5 \times (10 \times 60) \text{ C}$$
$$Q = 3000 \text{ C}$$
$$W = Q \times V = 3000 \times 12$$
$$W = 36\,000 \text{ J or } 3.6 \text{ kJ}$$

Electromotive force can be produced through:

- a chemical source
- heat
- electromagnetic induction (see page 50).

Chemical sources

When two different metals are placed in an **electrolyte**, ions are drawn towards one metal and electrons to the other. This is called a cell and produces electricity. A set of several cells joined together is called a battery.

Electrolyte

A chemical solution that contains many ions. Examples include salty water and lemon juice. In major battery production, these may be alkaline or acid solutions, or gels.

Heat

Simply wrapping a copper wire around a nail and heating one side with a flame can produce electricity, although in very small amounts. This is known as thermoelectric generation. Because the two metals react to the differences in temperature on the heated side and the cool side, a magnetic effect occurs (see magnetism, later in this unit), which creates a current and emf. This process is sometimes called the Seebeck effect. This principle is used in thermocouples, which are used to sense temperature. The amount of electricity generated is in proportion to the temperature.

In some waste disposal plants where waste is burnt, this effect is used to generate electricity.

ACTIVITY

Find a lemon, a zinc-coated nail and a piece of copper. Place the nail and copper into the lemon at opposite ends, ensuring there is a good gap between them inside the lemon. Use a sensitive voltmeter to measure the voltage between the two metals. You can see that a cell has been produced.

WAVEFORM

The characteristic of a.c. is its oscillating waveform, referred to as a 'sine wave'. This oscillating current arises through the changing position of the winding relative to the magnetic field within the alternator generating the current.

In practice, the a.c. supply at a workplace is unlikely to have a waveform that is a pure sine wave because of harmonics (contamination) arising from the connected loads. This will result in distortion of the waveform shape. The shape of the waveform of an a.c. supply may be displayed on an oscilloscope.

Because of its oscillating nature, alternating current and its associated voltage have additional factors that must be considered in describing the nature of the supply. These include the frequency of oscillation and the 'effective' current and voltage, whose values are continuously changing with time.

The output of an a.c. system, when measured and tracked, is usually referred to as a waveform.

As the rotating machine induces voltage, the value rises to a peak, falls to zero, then falls to a peak negative value, and then rises back to zero.

This single waveform represents one full turn of the alternator.

In the diagram the sinusoidal waveform is produced as the rotating conductor cuts the magnetic field set up by the permanent magnet. As the conductor is rotated at a constant speed and the number of conductors is fixed, the only variable is the amount of magnetic flux being cut, which can change the value of emf induced. This is represented by the waveform.

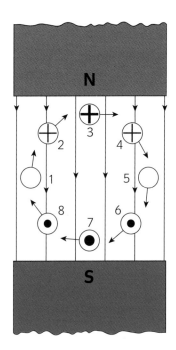
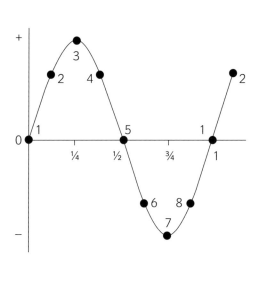

The emf generated per rotation: the time taken for the cycle to return to its starting point is periodic time (T), shown on the waveform from the first position 1 to the second position 1

As the conductor in the diagram above starts at position 1, the direction of motion is in line with the field. Therefore no flux is crossed and no emf is induced.

As the conductor rotates, the number of lines of flux increases together with the angle at which the conductor cuts the flux lines, until the maximum value of emf is produced at position 3. As the conductor moves from position 3 to 5, the emf decreases to zero, at which point the conductor cuts no magnetic flux lines because it moves in the same direction as the flux. As the conductor rotates from position 5 to 7, the induced emf increases, reaching its peak value because the conductor is cutting the maximum number of flux lines at right angles. This time the polarity is reversed, and reaches the negative peak value. As the conductor rotates back to position 1, the number of flux lines cut is reduced, so that the output decreases to zero.

This is the complete cycle, producing a sinusoidal waveform (or sine wave) with zero at the start, the mid-point and the end.

This waveform is a sine wave in which the instantaneous voltage v (the voltage at any one point) can therefore be calculated as:

$$v = V_{max} \times \sin\theta$$

where:

V_{max} = the maximum induced voltage in the coil

$\sin\theta$ = the sine of the angle at that point.

The value V = can be plotted for any angle, as shown in the table.

> **KEY POINT**
>
> As the conductor rotates one full circle, it rotates 360°, so the sine wave is a representation of a circle over a period of time.

Coil angle (θ)	0	30	45	60	90	135	180	225	270	315	360
$v = V_{max} \sin\theta$ where V_{max} = **100 V**	0	+50	+70.71	+86.67	+100	+70.71	0	−70.71	−100	−70.71	0

The values can be plotted for any angle to give a value for the instantaneous value based on the value of V_{max}.

Frequency

Using UK frequencies, a.c. values change at a rate of 50 Hz, which means that the cycle repeats 50 times every second. This means that the voltage and current waveforms are rising to maximum positive and maximum negative and back to zero every 0.02 seconds. Because of this, the calculation to determine values is slightly more complex than that for d.c. calculations.

The time taken for the cycle to return to its starting position is the periodic time (t). The number of cycles per second is called the frequency (f), measured in hertz (Hz). Frequency is defined as:

$$f = \frac{1}{t}$$

where:

t = periodic time, in seconds (s)

f = frequency, in hertz (Hz)(cycles per second).

In the UK, the electricity supply frequency is 50 Hz. Therefore the time (t) to complete one cycle is:

$$t = \frac{1}{f} = \frac{1}{50} = 0.02 \text{ s}$$

Therefore the time taken to complete one cycle of the UK electricity supply is 0.02 seconds.

As the values change, different values of voltage and current will be obtained, depending on the actual time and the position on the waveform the instant that the value is measured. The symbols for these values are represented by small letters: instantaneous voltage (v) and instantaneous current (i).

> **ACTIVITY**
>
> The standard a.c. frequency in the UK is 50 Hz. Find out the frequencies used in other countries and the reasons for any differences that you find.

5.3 Explain how characteristics of a sine-wave affect the values of a.c. voltage and current

Mean

The mean is the average of the numbers; a calculated 'central' value of a set of numbers. To calculate the mean, just add up all the numbers of the set, then divide by how many numbers there are in the set.

Root mean square (RMS)

The square root of the mean of the squares of the value.

MAXIMUM OR PEAK VOLTAGE AND CURRENT

The **mean** voltage (V_{av}) or mean current (I_{av}) of the waveform is zero because half a cycle is positive and the other half is negative, cancelling each other. However, to obtain the mean voltage or current of any half cycle it is necessary to multiply the peak voltage (V_{max}) or peak current (I_{max}) by 0.637. This is normally calculated by dividing a half waveform into enough equal divisions to be accurate but not too cumbersome. The values are added up and divided by the number of values taken to give the mean value (ie 0.637 peak for a sine wave).

The **root mean square** (RMS) value of a waveform is the equivalent value of a.c. that provides the same heat or work as a d.c. output over the same time period. To find the RMS value, each individual value is squared, these squares are added together and then the total is divided by the number of values to give the mean square. The square root of this value gives the RMS, which for a sine wave is 0.707 of the peak value.

Values of a.c. are taken to be RMS values unless otherwise specified because that is the effective current and the same relationship applies to the voltage associated with the a.c. that has the same waveform. When we consider the supply voltage to a house in the UK as 230 V, this is the RMS value. The actual peak value is 325 V. Peak values are relevant in some circumstances, eg concerning the specification of cable insulation which must be capable of withstanding the peak voltage.

Peak-to-peak value

The peak-to-peak value is the measurement between the positive peak and the negative peak on a cycle. This equates to twice the peak value of any half cycle, assuming that the centre of the waveform is based at zero.

What is the peak voltage if the RMS voltage is:

a) 110 V
b) 240 V
c) 400 V?

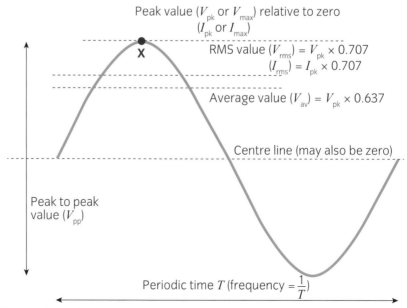

Peak value (V_{pk} or V_{max}) relative to zero (I_{pk} or I_{max})

RMS value (V_{rms}) = V_{pk} × 0.707
(I_{rms}) = I_{pk} × 0.707

Average value (V_{av}) = V_{pk} × 0.637

Centre line (may also be zero)

Peak to peak value (V_{pp})

Periodic time T (frequency = $\frac{1}{T}$)

Different values of a sine wave

THE CHANGING POWER SYSTEM

The industry for the supply of electricity in the UK has been undergoing changes since the power sector was privatised in the 1990s. Before privatisation, 14 area electricity boards (England and Scotland) provided the supply and distribution of electricity to customers, and the Central Electricity Generating Board (CEGB) owned, maintained and operated the power stations and main transmission lines. Each electricity board operated its own distribution network and supplied power to the customer, be that domestic industrial or commercial.

Large power stations operating on coal, natural gas and nuclear power were connected directly to the transmission system. The majority of the coal and gas-fired stations were located in the north of the UK along the spine of the country with the nuclear sites located by the coast.

Assessment criteria

6.1 Describe how electricity is generated and transmitted for domestic and industrial/commercial consumption

SmartScreen Unit 309
PowerPoint 6 and Handout 6

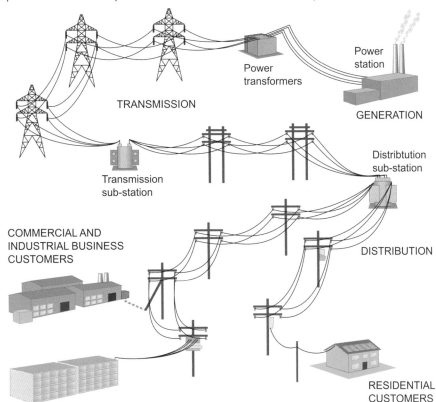

Electricity generation, transmission, distribution and use

Upon privatisation in 1990, the 12 regional English electricity boards were turned into 12 separate companies with the intention of them buying their electricity wholesale, at market rates, and mainly from the three concerns that produced the electricity: National Power and Powergen, which took over the CEGB's big coal-fired power stations in 1991, and British Energy, owner of the newest nuclear stations. The transportation of electricity between power stations and from region to region, was carried out by National Grid, owned jointly at first by the 12 local electricity firms and then after 1996, as an independent commercial enterprise.

However, in addition to the large generating stations connected to the transmission system, an increasing number of a small electricity generating plants are connected throughout the distribution networks rather than the transmission system. Generation connected to the distribution network which is owned by the Distribution Network Operators (DNOs) is called Distributed Generation (DG). This results in power flowing from both the distribution network to customers and from customers with DG into the distribution network.

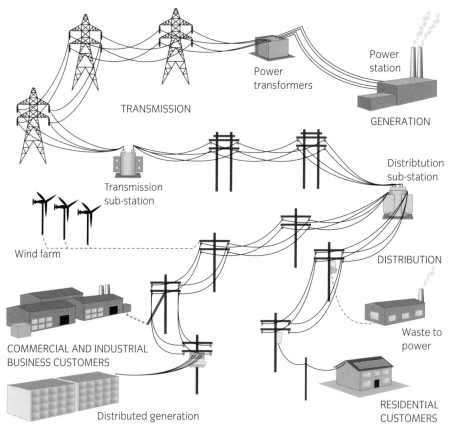

Electricity generation, transmission, distribution and use with embedded generation

The different organisations

Generators – generating organisations own, operate and maintain power stations that generate electricity from different energy sources: coal, gas, hydro and nuclear. As new technologies emerge, the mix of generation is being augmented with wind, solar, wave, pumped storage and tidal power.

Transmission System Owner – there are three transmission licence holders in Great Britain: National Grid, Scottish Power and SSE. They own and maintain the high voltage transmission system.

System Operator – because it is difficult to store large quantities of electricity, the demand has to be balanced with the generation output, and the System Operator can ask generators to increase output or large customers could be asked to reduce their demand (subject to contract conditions).

Distribution Network Operator (DNO) – a DNO owns, operates and maintains a public distribution network. There are presently seven DNOs in Great Britain and they often form part of a group that undertakes other areas of business, such as electricity supply.

Independent Distribution Network Operator (IDNO) – designs, builds, owns and operates and maintains a distribution network which is normally an extension of an existing DNO network. IDNOs build networks for new developments such as business parks, retail and residential areas.

Suppliers – supply is the retail arm for the provision of electricity. Suppliers buy in bulk and then sell to customers. They are responsible for providing bills and customer services and arranging metering.

THREE-PHASE A.C. GENERATORS

In the UK, three-phase generators are used in power stations.

Large-scale a.c. generation

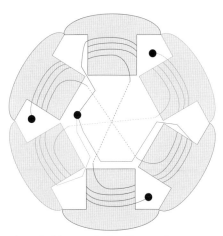

Six-pole (three pairs) salient pole rotor (star wound) for an alternator

A three-phase a.c. generator has a stator with three sets of windings arranged so that there is a phase displacement of 120°. The three-phase output is produced by either star- or delta-connected windings on the stator.

How a.c. generators work

For simplicity, the description below relates to one phase. However, it is important to remember that there are three phases, displaced at 120° from each other.

As each pair of poles passes through the strongest part of the magnetic field at right angles, the maximum electromotive force (emf) is induced into that particular phase. At that point, the other two pairs are in a weaker part of the field and a lower voltage is induced. The moving rotor is connected to the stationary external connections by brushes and slip rings, which keep each phase in constant contact.

The output of an a.c. system, when measured and tracked, is usually referred to as a waveform. This is because, as the rotating machine induces an emf, the value rises to a peak, falls to zero, then to a negative peak value and then rises back to zero.

ASSESSMENT GUIDANCE

Remember, $E = \beta l v$

ASSESSMENT GUIDANCE

Rotor bars are also formed by casting aluminium into the rotor.

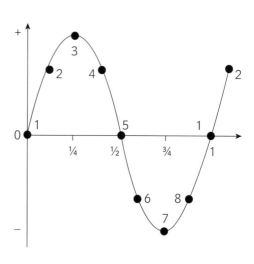

The emf-generated per phase per rotation

The time taken for the cycle to return to its starting position (from position 1 back to 1 in the example above) is the periodic time t. This process can be described in terms of Faraday's law because the rotation of the coil continually changes the magnetic flux through the coil and therefore generates an emf.

In the UK, large amounts of electricity are generated at high voltage in power stations. This is typically between 23 and 25 kV and is transformed to EHV (extra high voltage) 275 kV or 400 kV systems through **step-up transformers**. Once the electricity is transmitted to its region, it is transformed down to a more manageable voltage through **step-down transformers**. These distribution systems then deliver electricity at the correct voltage for the load, usually ending with an 11 000 V or 400 V transformer to supply both three- and single-phase installations at a local level of 230 V or 400 V.

A network of circuits, overhead lines, underground cables and substations link the power stations and allow large amounts of electricity to be transmitted around the country to meet the demand. Alongside the seven local distribution networks operating at 132 kV, 66 kV, 33 kV and 11 kV, there are also four high voltage transmission networks operating at 400 kV (super grid) and 275 kV (the grid) in the UK. The 400 kV network was installed in the 1960s to strengthen the 275 kV system which began operating in 1953. The 400 kV network has three times the power carrying capacity of one 275 kV line and eighteen times the capacity of a 132 kV line.

Primary distribution by the DNOs is usually carried out at 132 kV, using double circuit steel tower lines feeding primary substations, which in turn feed supplies at either 66 kV or 33kV. The purpose of these primary substations is to supply larger industrial installations and the secondary distribution networks in urban and rural areas. Secondary distribution networks carry supplies from the primary substations via overhead lines on wooden poles or underground cables. Customers are connected at low voltage 230–400 V single or three phase.

Step-up transformer

A transformer that has a proportionally higher number of turns on the secondary (output stage) than on the primary (input stage).

Step-down transformer

A transformer that has a proportionally higher number of turns on the primary than on the secondary.

Interconnected capacity

Interconnectors between Europe and the UK provide a pooling of capacity and diversity of supplies between the UK and Europe. This is provided by High Voltage Direct Current which is a way of conveying electricity over very long distances with fewer transmission losses than an equivalent HVAC solution. It also provides greater control over the transmission of electricity, including ability to change size and direction of power flow. Direct current supplies are usually obtained from a.c. mains supplies, first by using a transformer to change to the required voltage and secondly by using a rectifier to convert the a.c. supply to d.c. Unfortunately, the rectification process is not perfect and some superimposed ripple is likely to appear on the d.c. output.

The current interconnectors in operation are:

- 2 GW between England and France
- 1 GW between England and the Netherlands
- 500 MW between England and Northern Ireland.

ASSESSMENT GUIDANCE

A small number of power stations burn household refuse, which to a certain extent solves the landfill problem.

Standby supplies have diesel generators for use when the public supplies are not available.

ACTIVITY

Why are hydroelectric systems normally confined to Scotland and Wales?

ACTIVITY

Dinorwig is an example of a pumped storage electricity scheme. Find out how and when Dinorwig produces electricity.

OTHER METHODS OF GENERATION

The majority of electricity generation is produced by the conversion of heat or thermal energy (steam) to some form of mechanical energy (ie by turning a turbine) which in turn forces a generator to turn.

The burning of fossil fuels (coal, gas, petroleum/diesel) and nuclear fission have been the main sources of the heat to produce the steam to drive turbines.

Steam to drive turbines can also be produced by the following:

- **biomass** – such as wood, palm oil and willow
- **solar thermal** – Sun's heat energy is transferred to fluids
- **geothermal** – either directly from steam in the ground or via heat transfer.

Other means of turning a turbine are described below.

- **Water** – hydroelectric, pumped storage or micro hydro systems where the water turns turbine blades. Unlike large-scale hydroelectric schemes, such as Three Gorges Dam in China, micro hydro systems use small rivers or streams to produce up to 100kW of power.

Dinorwig (North Wales) pumped storage electricity generation scheme

- **Wind** – the turbines used in a wind farm for commercial electricity generation are usually three bladed and can have a wing tip speed of 200 mph.

A typical off shore wind farm installation

A small proportion of the UK's electricity is already generated from renewable sources but, with concerns about the depletion of fossil fuels, this is expected to grow significantly in the next few years. The Government is targeting a 34% reduction in carbon emissions by 2020, and 80% by 2050.

Other ways that electricity can be generated are:

- **solar photovoltaic cells (PV)** – these convert sunlight into electricity and although the PV concept is associated with small PV panels on domestic and commercial premises, many large PV power stations have been installed with hundreds of MW being produced

- **combined heat and power (CHP)** including micro CHP – in conventional power stations the waste heat is normally discarded via large cooling towers, whereas in a CHP generation system the waste heat or thermal energy is captured and used for heating schemes or production processes

- **batteries and cells** – When two different metals are placed in an **electrolyte**, ions are drawn towards one metal and electrons to the other. This is called a cell and produces electricity. Many cells joined together are called batteries.

Electrolyte

A chemical solution that contains many ions. Examples include salty water and lemon juice. In major battery production, these may be alkaline or acid solutions, or gels.

6.4 Describe the main characteristics of:

- single-phase electrical supplies
- three-phase electrical supplies
- three-phase and neutral supplies
- earth fault loop path
- star and delta connections

SOURCE AND ARRANGEMENTS OF SUPPLY

As was discussed earlier, a number of different electricity supply systems may be used in work premises, catering for specific requirements. These may be d.c. or a.c. operating at different voltages, and in the case of a.c. the supply may be single- or three-phase.

Direct current is not used for public electricity supplies (with the exception of the links between England, Netherlands, Ireland and France) but has some work applications, eg battery-operated works plant (fork lift trucks, etc). Certain parts of the UK railway system, in particular the London Underground and services in Southern England, also use d.c. for traction supplies.

Alternating current is the distribution system of choice for electricity suppliers all over the world. The main reason is its versatility. Using an a.c. supply permits wider scope concerning circuit arrangements and supply voltages, through the use of transformers, which enables the supply voltage to be changed up or down.

These different types of a.c. supply systems will now be explained.

ASSESSMENT GUIDANCE

In very rare circumstances an installation may be supplied with a two-phase and neutral supply. This is not to supply 400 V equipment but perhaps a large heating load which cannot be accommodated by a single-phase supply only.

Single-phase and neutral a.c. supplies

As with d.c. the simplest a.c. supply arrangement is a two-wire system, known as a single-phase supply. This is the arrangement provided for domestic premises, as well as for many supplies within work premises (lighting, socket outlets, etc). The two conductors are referred to as the line conductor (L) and the neutral conductor (N). The neutral conductor is connected to earth at every distribution substation on the public supply system and therefore the voltage of the neutral conductor, with reference to earth, at any point should be no more than a few volts. The voltage between the L and N conductors corresponds to the nominal supply voltage (230 V).

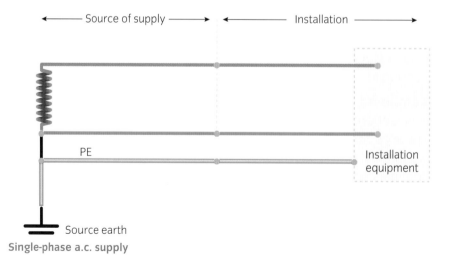

Single-phase a.c. supply

Balancing a single-phase load across a three-line supply

In the UK, electricity is provided from a delta star transformer (commonly referred to as a DYN11) with an earthed star point on the secondary side of the transformer. This is normally distributed as a three-phase four-wire system, as indicated below.

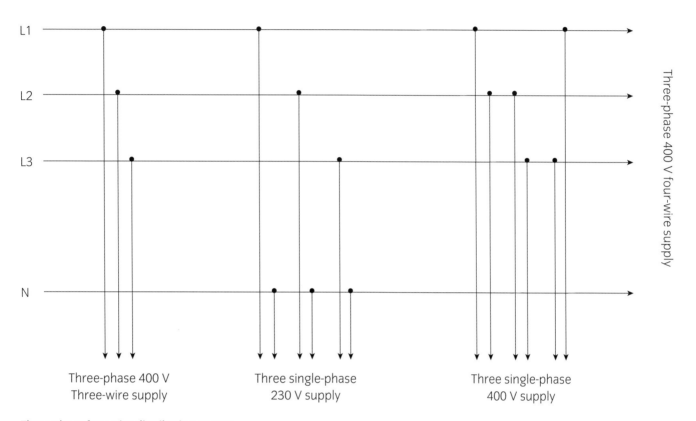

Three-phase four-wire distribution system

Balanced three-phase systems as described above will balance out, therefore giving no neutral current.

Single-phase loads require a neutral connection. With single phase loads, the neutral will carry any out-of-balance current. This should be kept to a minimum to reduce cable and switchgear sizes and to ensure maximum transformer use. Any difference in load will be 'mopped up' by current flowing in the neutral conductor.

It must be mentioned at this point that the effects of neutral current become much more complex with different loads. This subject is for later on in your studies.

Values of neutral current can be determined simply by using a scale drawing based on an equilateral triangle. If all phases are balanced, therefore equal, all three sides of the triangle will meet. If they are not balanced, there will be a gap, which represents the neutral current.

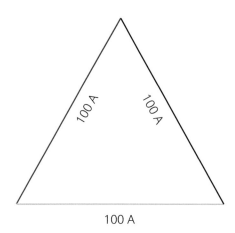

This triangle represents a balanced system

ACTIVITY

The load will always be unbalanced unless naturally balanced equipment is used. Identify some electrical equipment that will maintain a balanced load.

If an unbalanced system had the following values:

- L1 – 70 A
- L2 – 100 A
- L3 – 60 A

the neutral current would be as shown in the following diagram.

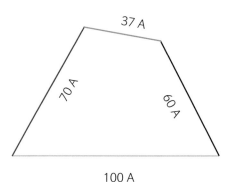

This diagram represents an unbalanced system

Represent the largest current by the base of the triangle. Draw the other two sides, each at 60° to the base, and mark off lengths to represent the other two currents, to scale. Join the top points of these two sides. This line represents the 'gap' and can be measured and converted, using the same scale. In this example, the gap, indicating the neutral current, represents approximately 37 A.

Three-phase and neutral a.c. supplies (Star connected)

Whilst single-phase a.c. supplies are adequate for domestic premises, the much higher loads typical of industrial and commercial premises would result in the need to use very large conductors to carry the high currents involved. The high currents would also give rise to a large volt drop.

However, it is possible to use a multi-phase arrangement, which effectively combines several single-phase supplies. If three coils spaced 120 degrees apart are rotated in a uniform magnetic field, we have an elementary system which will provide a symmetrical three-phase supply.

The usual arrangement is a three-phase system employing four conductors – three separate line conductors and a common neutral conductor, as shown in the following diagram. Most sub-station transformers in the distribution system that delivers power to houses are wound in a delta-to-star configuration. A neutral point is created on the star side of the transformer.

The three line conductors (L_1, L_2 and L_3) were previously distinguished by standard colour markings: red, yellow and blue. However, European harmonisation has now resulted in these conductor colours being changed to brown, black and grey.

ACTIVITY

Why is the star point of the transformer connected to earth?

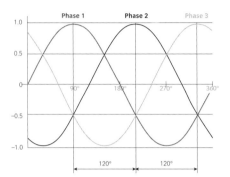

Three-phase sine wave where each waveform is 120° apart

ASSESSMENT GUIDANCE

If you look at the three-phase waveform you will see that all three currents are constantly changing. At 90°, L_1 shown in brown is 100% positive and L_2 shown in black and L_3 shown in grey are 50% each negative. As you move along you will see that the +ve values always equal the negative value.

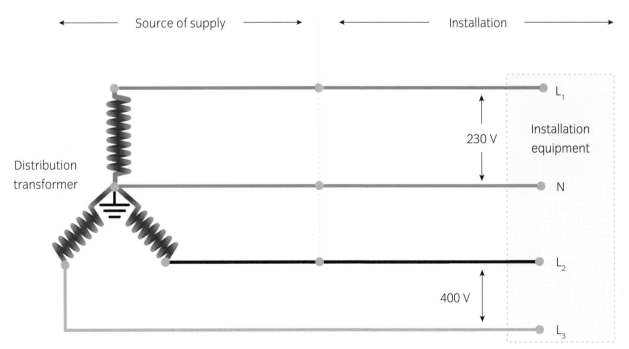

Three-phase a.c. supply

VOLTAGE AND CURRENT IN STAR-CONFIGURED SYSTEMS

In a star (Y) connected load:

- the line current (I_L) flows through the cable supplying each load
- the phase current (I_P) is the current flowing through each load.

So:

$$I_L = I_P$$

and:

- the voltage between any line conductors is the line voltage (V_L)
- the voltage across any one load is the phase voltage (V_P)

so:

$$V_P = \frac{V_L}{\sqrt{3}} \text{ or } V_L = V_P \times \sqrt{3}$$

In a balanced three-phase system there is no need to have a star-point connection to neutral as the current drawn by any one phase is taken out equally by the other two. Therefore the star point is naturally at zero.

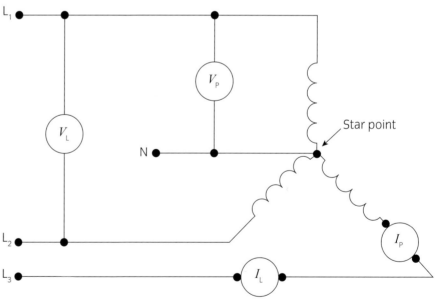

Star-connected load

So if a line current is 10 A, the phase current will also be 10 A. If the line voltage was 400 V, the phase voltage would be:

$$\frac{400}{\sqrt{3}} = 230 \text{ V}$$

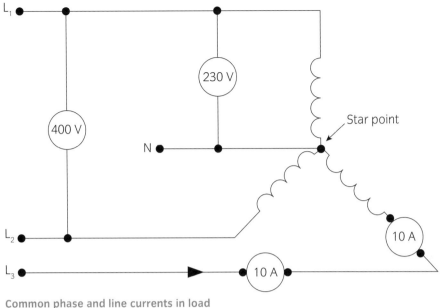

Common phase and line currents in load

Neutral current in three-phase and neutral supplies

In a balanced three-phase system there is no requirement to have a star-point connection as the three phases have a cancellation effect on each other. Therefore the star point is naturally at zero current. While the load is balanced and the waveforms are symmetrical (not containing harmonics or other waveform distorting influences), this statement is relatively accurate. However, in practice, this is sometimes not the case.

Where the load is not in balance, different currents circulate in the load through the source winding and back. This gives rise to a change in star-point voltage, which can result in the system 'floating' away from its earthed reference point. In essence, a current will flow in the neutral. The three ways in which this current value could be determined are:

- by phasor, an accurate method that indicates the angle at which the maximum current occurs
- by calculation, which gives an accurate value
- by equilateral triangle, which gives a good indication of the value.

Three-phase a.c. supply (Delta connected)

In a delta-connected system, each of the three coil windings is wound in such a way that each winding has a start and finish and the start of each winding is connected to the finish of the other winding. This gives a triangle or delta configuration. The supply is then taken from the interconnection of each winding which means that a three phase system can be obtained using only three wires.

Voltage and current in delta-configured systems

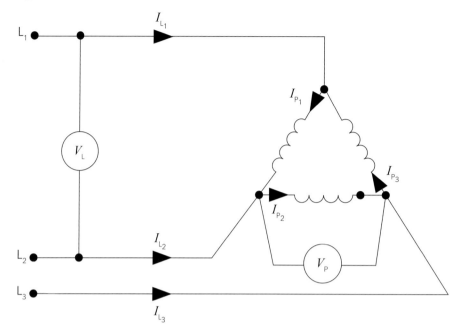

Delta-connected load similar to delta-connected supply

In a delta (Δ) connected load:

- the line current (I_L) flows through the cable supplying each load
- the phase current (I_P) is the current flowing through each load.

So:

$$I_P = \frac{I_L}{\sqrt{3}} \text{ or } I_L = I_P \times \sqrt{3}$$

and:

- the voltage between any line conductors is the line voltage (V_L)
- the voltage across any one load is the phase voltage (V_P)

so:

$$V_L = V_P$$

As there is no provision for a neutral connection, items such as delta motors would automatically be balanced – but complex loads on transmission systems could be unbalanced.

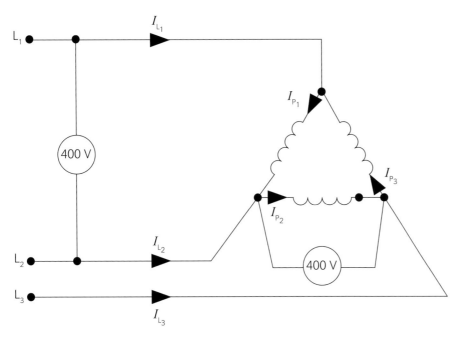

Phase and line currents in the load

Therefore if the phase current is 100 A and the phase voltage is 400 V, the line current can be calculated as follows:

$$I_{line} = 1.732 \times 100 = 173.2\,A$$

Earth fault loop impedance

Earth fault loop impedance is required to verify that there is an earth connection and that the value of the earth fault loop impedance is less than or equal to the value determined by the designer and which was used in the design calculations.

The earth fault current loop comprises the following elements, starting at the point of fault on the line to earth loop:

- the protective conductor
- the main earthing terminal and earthing conductor
- for TN systems (TN-S and TN-C-S), the metallic return path or, in the case of TT systems, the earth return path
- the path through the earthed neutral point of the transformer
- the source phase winding
- the line conductor from the source to the point of fault.

Earth fault loop impedance

The impedance of the earth fault current loop starting and ending at the point of earth fault. This impedance is denoted by the symbol Z_s.

ACTIVITY

Using an earth loop impedance tester to measure the earth electrode resistance involves disconnecting which conductor?

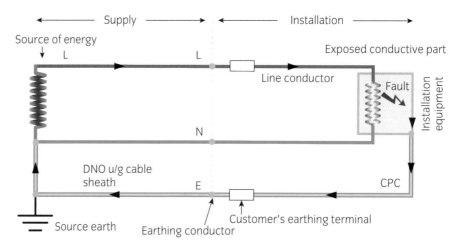

Earth fault loop

The earthing system adopted will determine the earth fault loop impedance, and this will determine the method of protection against electric shock.

- TN-C-S systems tend to have low earth fault loop impedances external to the installation, of the order of $0.35\,\Omega$.

- TN-S systems tend to have higher earth fault loop impedances compared to TN-C-S systems. The typical maximum declared value is $0.8\,\Omega$.

- When TT systems are adopted, the resistance of the installation earth will be high (of the order of a $100\,\Omega$). This means that residual current devices (RCDs) will need to be adopted for protection against electric shock as they operate at much lower earth fault currents than standard protective devices.

THE OPERATING PRINCIPLE OF TRANSFORMERS

The use of alternating current, rather than direct current, gives more scope for circuit and supply voltages as they can be changed up or down by the use of a transformer.

Transformers are fundamental to the safe and efficient use of electricity. They range from step-down transformers, which provide an extra-low voltage supply for small electrical appliances, to large step-up transformers that produce voltages of up to 400 kV for power transmission purposes. The wide range of sizes and capabilities corresponds to the everyday requirements of electricity usage.

Although there are many different types and sizes of transformer, they all operate on the same principle. The basic principle is that in two independent coils (windings), a change in the magnetic flux of one coil can induce a magnetic change in the second coil. This is known as static inductance. It is further enhanced when the two electrically separated coils are wound onto a common magnetic core, which creates a common magnetic circuit.

If an a.c. supply is applied to one coil (known as the primary), the magnetic flux produced by the first coil rotates and cuts through the second coil (known as the secondary), producing an emf in the secondary. This configuration works with a.c. or chopped d.c. The field in the first coil rises and falls rapidly, causing inductance in the adjacent coil.

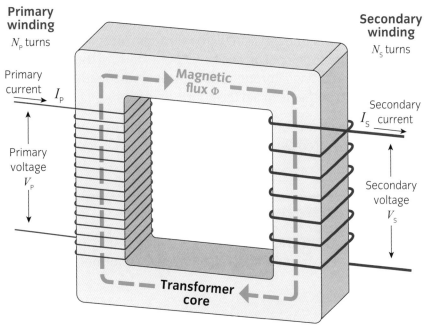

Primary winding N_P turns

Secondary winding N_S turns

Primary current I_P

Magnetic flux Φ

Secondary current I_S

Primary voltage V_P

Secondary voltage V_S

Transformer core

Simple transformer arrangement

The ease with which voltages can be changed up or down, coupled with the fact that it is more economical to transmit electricity over long distances at high voltage, is an important reason why a.c. has been adopted for mains electricity supplies.

Transformers are very efficient devices, particularly when working on full load and, since there are no moving parts, require little maintenance.

TRANSFORMERS

There are a wide range of electrical transformers all designed for different uses and purposes. The designs may differ but the fundamental principles are the same.

Power transformers

Power transformers are the most common type of transformer and are used as:

■ step up transformers in the transmission of electrical energy from power stations (25 kV–400 kV) and step down from 400 kV to 132 kV

■ step down transformers in distribution substations from 132 kV–11 kV, and 11 kV to 400 V for use in commercial and domestic situations

■ step down transformers to convert mains voltage 230 V to extra low voltage to power electronic equipment.

The construction of insulated laminations minimises the eddy current produced during the transformation process.

For large power transformers used at generating and distribution substations, oil is used as a coolant and insulating medium. For small rating transformers, the oil is circulated through ducts in the coil assembly and through cooling fins fixed to the body of the transformer tank or in higher ratings, through separate air-cooled radiators that may use pumps and fans to aid the cooling process.

Cast resin power transformers, where the windings are encased in epoxy resin, are often used where there is a fire risk in indoor situations.

Current transformers

Current transformers (CTs) have many uses. Those that involve metering require extremely accurate current transformers, such as Class X CTs.

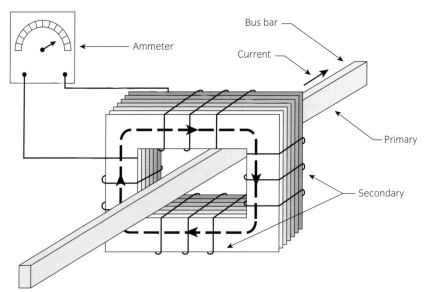

A current transformer with secondary winding around the primary conductor

A current transformer has a primary winding, a magnetic core and a secondary winding. The core and secondary winding surround the primary winding, which is a simple conductor giving one single turn.

The a.c. flowing in the primary conductor produces a magnetic field in the core, which then induces a current in the secondary winding circuit.

It is extremely important when positioning the current transformer to ensure that the primary and secondary circuits are efficiently coupled to give an accurate reading. It is also very important to ensure that the current transformer is always connected on the secondary side, either by a measuring instrument or a shorting link. If the CT is left open circuit, a large voltage discharge will occur.

Isolation transformers

Isolation transformers are available for step-up, step-down or straightforward 1:1 isolation purposes. Their uses are quite diverse but ultimately they are intended to ensure electrical separation from the primary supply, usually for safety purposes.

These transformers have additional insulation and electrostatic shielding. In some cases, one or both sides of the transformer remains separated from earth by a resistance or they have no connection at all.

A common use for isolation transformers is the domestic shaver socket arrangement, which is isolated from earth to minimise any shock hazards from touching live parts. Because the parts are isolated and the current cannot flow back to its origin (the secondary side of the transformer) via the earth path, the shaver socket cannot deliver a shock under first-fault conditions. Isolation transformers are also used in many medical situations in order to prevent first-fault failures. These systems are also known as medical isolated power supplies (IPS).

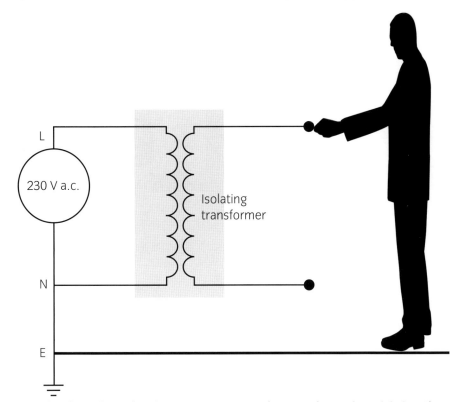

An isolated supply, such as in a BS EN 61558-2-5 shaver socket outlet, minimises the risk of shock

ASSESSMENT GUIDANCE

Remember, you will still get a shock between the two transformer output terminals.

Voltage transformers

A voltage or potential transformer has two windings wound around a common core. See the diagram of the simple transformer arrangement on page 77.

Earthing transformers

Earthing transformers are used to provide a physical neutral for power transformers with a delta-connected secondary. The earthing transformer neutral point has low impedance and is suitable for different earthing systems such as solid earthing, NER earthing (resistor) and resonance earthing (Arc Suppression Coil).

KEY POINT

The lower voltage winding is normally wound closest to the core for safety reasons. Some windings are placed on top of each other; other types use a sandwich arrangement.

Transformer cores

The core of a transformer is generally one of two types: core or shell.

The shell-type transformer is regarded as being more efficient as the magnetic flux is able to circulate through two paths around the core. The core-type transformer only channels flux through one path, meaning that some of the flux is lost at the core corners (leakage flux).

ACTIVITY

There is a third type of core. Find out what it is.

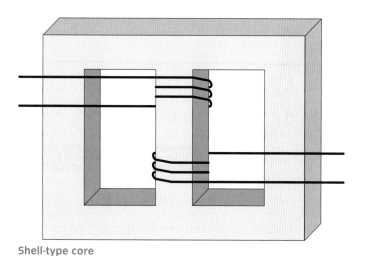

Shell-type core

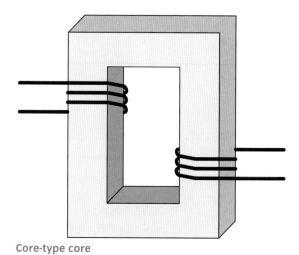

Core-type core

ASSESSMENT GUIDANCE

Eddy current and hysteresis are known as iron losses and, given a fixed frequency, are normally considered to be constant. Copper losses occur in the windings and are proportional to the square of the load current.

Transformer laminations

Transformer cores are made up of laminations. Laminations are thin, electrically insulated slices of metal that, when stacked on top of one another, form a large core. These thin laminations are used to reduce eddy currents in the transformer core.

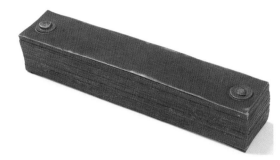

Laminations used to form a transformer core

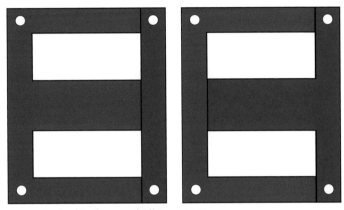

Laminations are stacked to form a shell-type core

Eddy currents are products of induction. Small rotating currents, like those swirling currents seen in rivers, rotate around a transformer core. If allowed to, these currents produce heat that ends up as a loss of energy. Making a transformer core with thin, insulated laminations prevents eddy currents from flowing, thus reducing any loss. Eddy current loss is a form of iron loss, that is, losses in the core of a transformer.

Another type of loss is hysteresis loss. This is due to the transformer core being magnetised in one direction, then re-magnetised in the other direction as the supply alternates. If the material requires energy to re-magnetise it, this also becomes a loss. Careful consideration of the material used to construct the core reduces hysteresis loss.

Further losses in transformers are copper losses. This is a loss due to the heating effect of current passing through the windings. The windings are resistances and as in any resistance, the power dissipated, or lost, is usually determined as I^2R.

This is because power is determined by:

$$P = V \times I$$

and voltage is determined by:

$$V = I \times R$$

so power can also be expressed as:

$$P = I \times I \times R \text{ or } I^2R$$

Inductance

There are two types of inductance in relation to transformers: self-inductance and mutual-inductance.

Self-inductance is where a transformer winding induces a magnetic field that rotates around the core and induces a current back into the winding, limiting current flow. This principle is used in choke/ballast units in fluorescent luminaires.

Mutual-inductance is where a primary winding induces a magnetic field that induces a current into a second winding, as in current and voltage transformers.

ASSESSMENT GUIDANCE

When working out transformer ratios, use the winding to winding values – that is, phase voltage to phase voltage (Vp – Vs).

Assessment criteria

6.7 Determine by calculation and measurement:

- primary and secondary voltages
- primary and secondary current
- kVA rating of a transformer.

RELATIONSHIP OF EMF PRODUCED AND NUMBER OF TURNS

The proportion of emf produced is related to the number of turns on the primary coil with respect to the secondary coil. This relationship can be determined by:

$$\frac{E_1}{E_2} = \frac{N_1}{N_2}$$

where:

E_1 = emf induced in the primary (V_1)

E_2 = emf induced in the secondary (V_2)

N_1 = number of primary turns

N_2 = number of secondary turns.

As power transformers have very low impedance, there is negligible error in assuming that the voltage on the primary is the same as the emf induced on the primary and likewise for the voltage and emf on the secondary.

A reasonable assumption is that:

$$\frac{E_1}{E_2} = \frac{V_1}{V_2}$$

Therefore:

$$\frac{V_1}{V_2} = \frac{N_1}{N_2}$$

The ratio of primary turns to secondary is referred to as the turns ratio, represented by:

$$\frac{N_1}{N_2}$$

Current is also affected by the turns ratio of a transformer, but in reverse to voltage, so:

$$\frac{V_1}{V_2} = \frac{N_1}{N_2} = \frac{I_2}{I_1}$$

Example

A transformer is wound with 560 turns on the primary and 20 turns on the secondary. Suppose that 230 V a.c. is applied to the primary winding. Calculate the output voltage.

$$\frac{V_1}{V_2} = \frac{N_1}{N_2}$$

Then:

$$\frac{230}{V_2} = \frac{560}{20}$$

So:

$$V_2 = \frac{230 \times 20}{560} = 8.21 \text{ V}$$

If the secondary current is 4 A, calculate the input primary current.

$$\frac{N_1}{N_2} = \frac{I_2}{I_1}$$

Then:

$$\frac{560}{20} = \frac{4}{I_1}$$

So:

$$I_1 = \frac{4 \times 20}{560} = 0.14 \text{ A}$$

ACTIVITY

Try these calculations with different combinations of primary and secondary turns or voltages.

Transformer power ratings

As a transformer is not a load, simply a method of changing voltage and current, it is rated in kVA. This is because certain loads may include a power factor, which can increase the current demand by the load. As a result, the rating of the transformer is given in volt amperes (VA) instead of kilowatts (kW). So when selecting a transformer for a particular load, the actual current drawn by the load must be multiplied by the voltage.

Understand how different electrical properties can affect electrical circuits, systems and equipment

Assessment criteria

7.1 Explain the relationship between resistance, inductance, capacitance and impedance

THE RELATIONSHIP BETWEEN RESISTANCE, INDUCTANCE, CAPACITANCE AND IMPEDANCE

Different components have different effects on a.c. circuits; these effects vary with frequency, depending on the components in the circuit. Other factors, such as resistance, have no effect on the waveform except to resist current flow.

Resistance

When a resistor is used in an a.c. circuit, the voltage drop and the current through the resistor are in phase, as the resistor has no effect on the circuit voltage or current, other than to restrict current flow proportionally to the voltage. The sinusoidal voltages and currents have no phase shift and are described as being 'in phase'. This in-phase or *unity* relationship can be understood using Ohm's law for the voltage and current:

$$V = I \times R$$

> **ASSESSMENT GUIDANCE**
>
> Ohm's law states that the current flowing in a circuit is proportional to the applied voltage and inversely proportional to the resistance.

As there is no phase shift effect, there is no requirement in a resistive circuit to include the power factor. Resistive loads in a.c. circuits include items such as incandescent lamps, water heater (immersion) elements and electric kettles. Any load that relies on current passing through a material and producing heat is resistive.

> **ASSESSMENT GUIDANCE**
>
> Power factor is the ratio of true power (watts) to volt-amperes:
>
> Pf = W/VA

The relationship between the voltage, current and resistance in an a.c. circuit are shown opposite with the circuit diagram and the phasor diagram and sine wave, together with the appropriate formula.

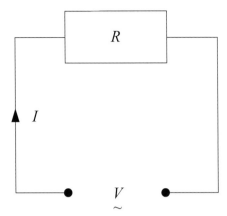

Relationship between voltage, current and resistance as shown by a circuit diagram

The circuit diagram shows the resistor connected to an a.c. supply.

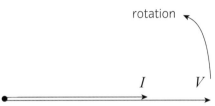

Relationship between the voltage and current as shown by a phasor diagram

The phasor diagram shows the relationship between the voltage and current. As the voltage and current are together, they are in phase or unity.

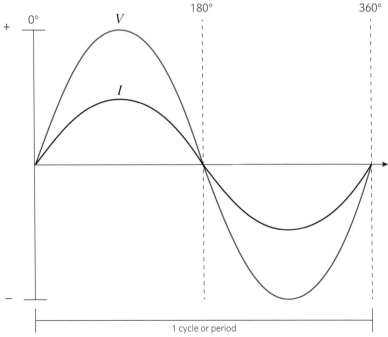

Relationship between voltage and current as shown by a wave form

The sine wave shows the voltage and current rising and falling at the same time. In a.c. circuits involving resistance only, Ohm's law applies as:

$$V = I \times R$$

Inductance

An inductor is a length of wire; sometimes this is a longer wire wound into a coil with a core of iron or air. An inductor is a component, such as a solenoid or field winding in a motor or a ballast unit in a fluorescent luminaire, that induces (produces) magnetism. Inductance is proportional to the inductor's opposition to a.c. current flow.

- The symbol for inductance is L.
- The unit of inductance is the henry (H).

As inductance increases and all other factors, such as frequency, remain the same, a.c. current flow reduces. This opposition to a.c. current flow is indicated as an increase in inductive reactance X_L.

We will assume that an inductor in a circuit produces pure inductance but this is actually impossible; as an inductor is a coiled wire, it also has resistance.

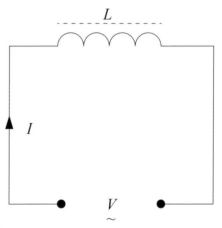

An inductor in a circuit

A pure inductance, as shown in the circuit, will create a phase shift that causes the current to lag behind the voltage by 90°.

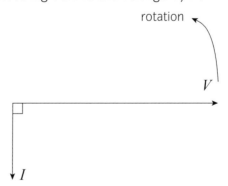

Phasor diagram showing phase shift

The phasor diagram shows the phase shift. When the phasor is rotated in the direction shown, the voltage leads the current or, as we normally say, the current *lags* behind the voltage. If the inductance was pure, this lag would be 90°. We shall explore how to create a phasor diagram later in this outcome.

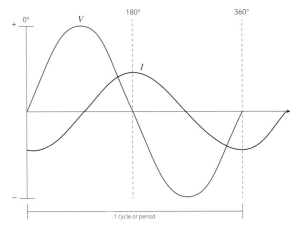

Sine wave diagram showing phase shift

If you look at the lagging current sine wave, you will see that the current does not begin its cycle until the voltage is 90° into its cycle. Where the voltage begins its negative half-cycle, the current is still in a positive cycle – meaning the two values are reacting to one another. This is known as reactance, X, measured in ohms (Ω). As the reactance is linked to an inductor, we call this inductive reactance (X_L).

The reactance in the circuit can be determined using the value of inductance (L) in henrys (H) and the frequency of the supply (f) measured in hertz (Hz). So

$$X_L = 2\pi fL$$

Example calculation

Determine the reactance when an inductance of 40 mH is connected to a 230 V 50 Hz supply.

As:

$$X_L = 2\pi fL$$

then:

$$X_L = 2\pi \times 50 \times 40 \times 10^{-3} = 12.56\,\Omega$$

If the inductance is pure, the reactance replaces resistance in Ohm's law so:

$$I = \frac{V}{X_L}$$

So using the example above, the value of circuit current would be:

$$I = \frac{230}{12.56} = 18.3\,\text{A}$$

Assuming the inductance to be pure, this current would lag the voltage by 90°. Remember, however, that the inductance cannot be pure as the coil or winding is made up of a conductor which has resistance. We shall study the effect of resistance on an inductor later in this outcome.

Capacitance

A capacitor is a device that is used to store and discharge energy. It does not contain a resistance so it does not dissipate energy. A capacitor *can* be pure. Capacitors are used in circuits for many reasons; most commonly, capacitors are found in fluorescent luminaires for power factor correction purposes. Capacitance (*C*) is measured in farads (F) – usually expressed in micro-farads. One micro-farad (µF) is one-millionth (10^{-6}) of a farad.

A capacitor has the opposite effect to an inductor when connected to an a.c. circuit; it causes the current to lead the voltage by 90°.

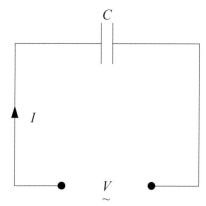

Circuit diagram with capacitor

The capacitor produces a reactance. This reactance, known as capacitive reactance (X_C), affects current flow.

Phasor diagram showing the effects of a capacitor

The phasor shows that the current leads the voltage by 90° in the direction of rotation.

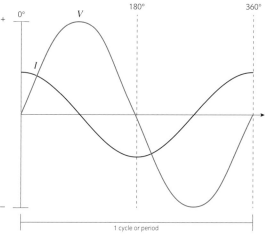

Sine wave diagram showing phase shift due to a capacitor

THE CITY & GUILDS TEXTBOOK

The sine wave diagram shows the current leading the voltage and, once again, the current and voltage are reacting. At various points in the cycle there will be reactance.

The capacitive reactance of a capacitive circuit can be determined, using:

$$X_C = \frac{1}{2\pi f C}$$

Example calculations

Determine the reactance when a 120 µF capacitor is connected to a 230 V 50 Hz supply.

As:

$$X_C = \frac{1}{2\pi f C}$$

then:

$$X_C = \frac{1}{2\pi \times 50 \times 120 \times 10^{-6}} = 26.52 \,\Omega$$

Where the capacitance is pure (that is no other components are connected in that part of the circuit), then X_C replaces resistance in Ohm's law.

Determine, using the example above, the current drawn by the capacitor.

So as:

$$I = \frac{V}{X_C}$$

then:

$$I = \frac{230}{26.52} = 8.6 \,A$$

Impedance

Impedance is the product of resistance and one or more of the other two components in a circuit. Where resistance and another component are connected in a circuit, the effect reduces the angle by which the current leads or lags. As resistance is in unity and the other components cause the current to lead or lag the voltage, and since more than one current cannot exist, the resulting current will fall in between, depending on the values of all the components.

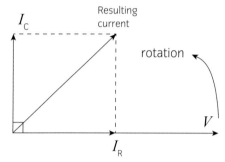

Phasor diagram showing a leading current due to a capacitor and resistor in the circuit

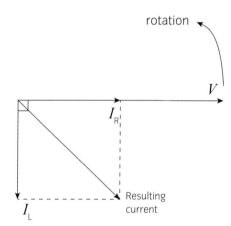

Phasor diagram showing the resultant lagging current due to the inductor and resistor in the circuit

This current is determined by using the circuit impedance (Z) measured in ohms. Impedance can be determined by:

$$Z = \sqrt{R^2 + X^2}$$

The value of reactance (X) depends on the component used so, if an inductance is connected with an impedance, X_L is used and for capacitance X_C is used.

Where a circuit contains both, the resulting reactance is used; this is the smallest value subtracted from the largest.

Example calculation

Determine the impedance if a circuit contained a 30 Ω resistance, an inductor having a 12.56 Ω reactance and a capacitor having a 26.52 Ω reactance.

The total reactance is:

$$X = X_C - X_L$$

as the capacitive reactance is the larger value.

So as:

$$Z = \sqrt{R^2 + X^2}$$

then:

$$Z = \sqrt{30^2 + (26.52 - 12.56)^2} = 33.08\,\Omega$$

Impedance triangle

The relationship between the different components in a circuit can be explored using an impedance triangle, as shown in the diagram.

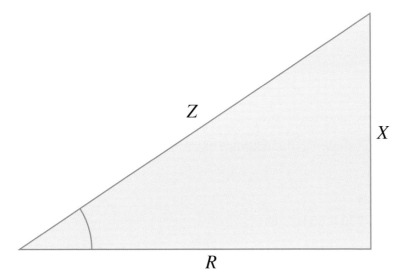

Impedance triangle

ASSESSMENT GUIDANCE

Make sure you draw the impedance triangle the right way round, depending on whether it is leading or lagging.

Using the triangle and Pythagoras' theorem, we can see that the impedance value alters depending on the values of resistance or impedance. Impedance triangles are drawn to scale. If the value of reactance increases or decreases, this affects the angle between the resistance (R) and impedance (Z). We shall explore this relationship later in this outcome when we look at power factor.

SmartScreen Unit 309

PowerPoint 7 and Handout 7

ACTIVITY

Calculate the current drawn by a 10 µF capacitor when connected to a 230 V 50 Hz a.c. supply.

CALCULATE UNKNOWN VALUES IN A.C. CIRCUITS

Components in a.c. circuits can be arranged in series, parallel or in combination. Each configuration has different effects on circuit properties.

Series connected circuits (RL series)

When an inductor is connected into a circuit, we consider that the inductor has pure inductance (pages 86–87). This, in reality, is not possible as the conductor that is used to form the inductor's coil has a resistance. These two properties are in series with one another. As with d.c. circuits, if components are connected in series, the current is constant (that is, it is the same through each component), but the voltage changes as it is 'lost' across each component, creating voltage drop or a potential difference.

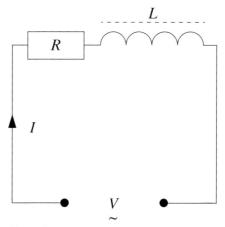

Components connected in series

The circuit shown in the diagram has the following values:

$R = 40 \, \Omega$

$L = 38 \, \text{mH}$

Supply = 230 V 50 Hz

Example calculation

Determine:

a) the inductive reactance (X_L)

b) the impedance (Z)

c) the total circuit current (I)

d) the value of voltage across each component (R_R) and (V_L).

a) As $X_L = 2\pi fL$ then $X_L = 2 \times \pi \times 50 \times 38 \times 10^{-3} = 11.93\,\Omega$

b) As $Z = \sqrt{R^2 + X^2}$ then $Z = \sqrt{40^2 + 11.93^2} = 41.74\,\Omega$

c) As $I = \dfrac{V}{Z}$ then $I = \dfrac{230}{41.74} = 5.51\,A$

d) The value of voltage across each component part is determined from Ohm's law:

$V_R = I \times R$ so $V_R = 5.51 \times 40 = 220.4\,V$

$V_L = I \times R_L$ so $V_L = 5.51 \times 11.93 = 65.73\,V$

As you can see, the two voltages do not add up to the supply voltage of 230 V. This is because the supply voltage would be the phasor sum of the two values as each component reacts differently to a.c. current.

We can determine the phasor sum in two ways, by calculation (using Pythagoras' theorem) or by constructing a phasor to represent the two components.

Determining the phasor sum by calculation

$$V_{supply} = \sqrt{V_R^2 + V_L^2}$$

so

$$V_{supply} = \sqrt{220.4^2 + 65.61^2} = 229.99\,V \text{ or } 230\,V$$

Constructing a phasor

As we have seen, a phasor diagram is a representation of the component values in an a.c. circuit and how the components lead or lag. Also, it shows the resulting supply characteristics.

Constructing a phasor; using a reference line

To construct a phasor, we must first decide on a reference line. In a series circuit the common component is current as it is the same throughout the circuit; in a parallel circuit, this would be voltage. With exception to the reference line, phasor diagrams must be drawn using a suitable scale.

Once a reference line is drawn, the line representing the voltage across the resistor is drawn. As this voltage is in unity to the current, this line is plotted on the reference line.

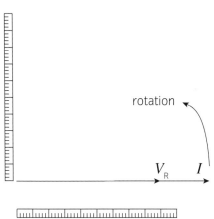

Drawing a phasor to scale

Then we add the value of the inductor voltage to the same scale.

As the current lags in an inductive circuit, the voltage V_L is drawn upwards. This is to show that the current (reference line) lags the voltage (given the direction of rotation). As we must assume this component to be pure, the line is drawn 90° from the reference line.

Then we construct a parallelogram from the two voltage values.

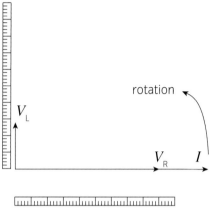

Drawing the voltage value

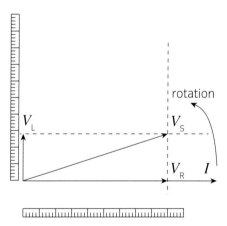

Constructing a parallelogram

Finally, the resulting supply voltage is drawn in, connecting the origin to the point where the two sides of the parallelogram intersect. The length of this line, to the scale used, represents the supply voltage V_S.

We can see from the phasor that the current lags behind the supply voltage by a particular angle less than 90° – the resistance acts against the pure inductance, meaning the current lag is not fully 90°.

Drawing in the supply voltage

RLC series circuits

If a circuit contains all three components, as in the circuit shown in the diagram, the components behave differently as the circuit current changes.

The circuit shown in the diagram has the following values:

$R = 40\,\Omega$

$L = 38$ mH

$C = 120\,\mu$F

Supply = 230 V 50 Hz

Example calculation

Determine:

a) the inductive reactance (X_L)

b) the capacitive reactance (X_C)

c) the impedance (Z)

d) the total circuit current (I)

e) the value of voltage across each component (R_R), (V_L) and (V_C).

a) As $X_L = 2\pi fL$ then $X_L = 2 \times \pi \times 50 \times 38 \times 10^{-3} = 11.93\,\Omega$

b) As $X_C = \dfrac{1}{2\pi fC}$ then $X_C = \dfrac{1}{2 \times \pi \times 50 \times 120 \times 10^{-6}} = 26.52\,\Omega$

c) As $Z = \sqrt{R^2 + X^2}$ and $X = X_C - X_L$ then

   $X = X_C - X_L = 26.52 - 11.93 = 14.59\,\Omega$

   So $Z = \sqrt{40^2 + 14.59^2} = 42.57\,\Omega$

d) As $I = \dfrac{V}{Z}$ then $I = \dfrac{230}{42.57} = 5.4$ A

e) The value of voltage across each component part is determined from Ohm's law:

   $V_R = I \times R$ so $V_R = 5.4 \times 40 = 216$ V

   $V_L = I \times X_L$ so $V_L = 5.4 \times 11.93 = 64.42$ V

   $V_C = I \times R_C$ so $V_C = 5.4 \times 26.52 = 143.2$ V

Once again, we could prove the circuit supply voltage by calculation or phasor.

ASSESSMENT GUIDANCE

In an RLC series circuit where the inductive and capacitive reactance cancel each other out, $R = Z$ and the power factor is 1 (unity). The current is V/R. High voltage may appear across the reactive components.

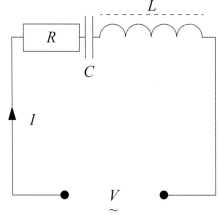

RLC circuit containing all three components

ACTIVITY

A capacitor and inductor each of 200 Ω reactance are connected in series to a 115 Ω resistor. What is the voltage across the capacitor if the supply voltage is 230 V?

Determining the phasor sum by calculation

$$V_{supply} = \sqrt{V_R^2 + V_X^2} \text{ where } V_X = V_C - V_L$$

Once again, subtract the smallest value from the largest:

$$V_X = 143.2 - 64.42 = 78.78\,V$$

So

$$V_{supply} = \sqrt{216^2 + 78.78^2} = 229.5 \text{ or } 230\,V$$

Constructing a phasor

We construct the phasor as before, but this time we insert the voltage across the capacitor before we construct the parallelogram. Remember, as the current leads the voltage in a capacitor, the voltage lags the reference line (in the downwards direction) by 90°.

Following this, we take the smallest value of voltage in the capacitor or inductor from the largest, just as in the calculation, so we end up with a resulting voltage V_X from which the parallelogram may be formed.

We can see that the current now leads the voltage by a particular angle as the capacitor is the stronger component.

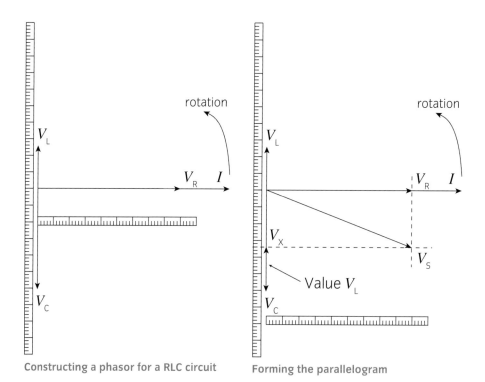

Constructing a phasor for a RLC circuit   Forming the parallelogram

Components in parallel

When components are connected in parallel, the voltage becomes the common component and current is split through each component.

The circuit shown in the diagram has the following values:

$R = 40\,\Omega$

$C = 120\,\mu F$

Supply = 230 V 50 Hz

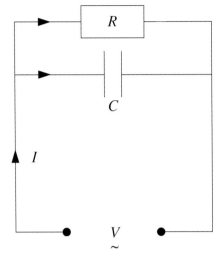

Example calculation

Determine:

a) the capacitive reactance X_C

b) the value of current through each component

c) the total circuit current.

I

Components in parallel

a) As $X_C = \dfrac{1}{2\pi fC}$ then $X_C = \dfrac{1}{2 \times \pi \times 50 \times 120 \times 10^{-6}} = 26.52\,\Omega$

b) $I_R = \dfrac{V}{R}$ so $I_R = \dfrac{230}{40} = 5.75\,A$

c) $I_C = \dfrac{V}{X_L}$ so $I_C = \dfrac{230}{26.52} = 8.67\,A$

In the same way as we did for voltage, we can prove the circuit supply current by calculation or phasor.

Proving the circuit supply current by calculation

$$I_{supply} = \sqrt{I_R^2 + I_X^2}$$

So

$$I_{supply} = \sqrt{5.75^2 + 8.67^2} = 10.4\,A$$

Constructing a phasor

Once again, drawn to a suitable scale, the phasor is used to determine the supply current. Notice that, this time, the voltage is the reference line as it is common to all components. As the capacitor is the component, the supply current ends up leading.

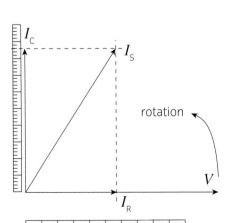

Constructing a phasor for a parallel circuit

Circuits with both series and parallel components

In practice, in electrical installations, electricians are more likely to come across circuits where the inductor and resistor are in series, such as in motor winding or in the choke/ballast in a fluorescent luminaire, and the capacitor is in parallel for power factor correction purposes.

In this situation, the current in the series section of the circuit is determined using impedance and the capacitor is determined using capacitive reactance.

The circuit shown in the diagram has the following values:

$R = 12\,\Omega$

$L = 88$ mH

$C = 150$ µF

Supply = 230 V 50 Hz

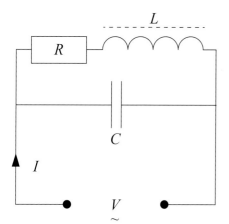

Circuit with components in both series and in parallel

Example calculation

First, we determine the inductive reactance:

$$X_L = 2\pi fL$$

so:

and:

$$X_L = 2\pi \times 50 \times 88 \times 10^{-3} = 27.64\,\Omega$$

so:

$$Z = \sqrt{R^2 + X_L^2}$$

$$Z = \sqrt{12^2 + 27.64^2} = 30.13\,\Omega$$

The current in the inductive/resistive section of the circuit is:

$$I = \frac{V}{Z}$$

so:

$$I = \frac{230}{30.13} = 7.63\,A$$

The current drawn by the capacitor is based on the capacitive reactance:

$$X_C = \frac{1}{2\pi fC}$$

so:

$$X_C = \frac{1}{2\pi \times 50 \times 150 \times 10^{-6}} = 21.22\,\Omega$$

therefore as:

$$I = \frac{V}{X_C}$$

then:

$$I = \frac{230}{21.22} = 10.83\,\text{A}$$

Showing this as a phasor to determine the total current requires a slightly different approach than before; we need first to construct a phasor using the series circuit values, then draw in the capacitance values. Before we do this, we need to understand how power factor affects the circuit and, therefore, the resulting phase angle.

POWER FACTOR AND THE POWER TRIANGLE

When the concept of power was explored at Level 2, the relationship between voltage and current was described as:

$$P = V \times I$$

If a load was purely resistive, this would be true. However, as many a.c. circuits have impedances and capacitors, the power behaves differently as an element of the power dissipated, due to the reactance of the circuit.

Power triangle

In power terms, this reactive part of the circuit (together with the true power relationship) can be explained using a power triangle.

As we can see, a resistive load gives the true power. The reactive (capacitive or inductive) load creates the reactive power element that affects the overall impedance. This is the apparent power.

Assessment criteria

7.3 Explain the relationship between kW, kVA$_r$, kVA and power factor

KEY POINT

Remember that, when an inductor or capacitor causes the current to lead or lag the voltage, at various points in the sine wave, the voltage and current are in opposition and, therefore, are reacting (giving reactance).

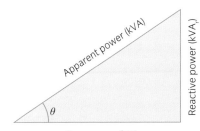

The true power relationship expressed in a power triangle

Therefore, having a reactive component in the load creates an apparent power. This draws more current from the load than if this was a purely resistive load. The apparent load is measured in volt-amperes (VA) or kilovolt-amperes (kVA). We can also see that an angle forms between the true power and apparent power. The cosine of this angle gives the power factor for the circuit or load.

So in order to determine the true power of a circuit we must apply this equation:

$$\text{true power (watts, W)} = V \times I \times \cos\theta$$

From this, we could also state:

$$\text{power factor } \cos\theta = \frac{\text{true power}}{\text{apparent power}} \text{ or } \frac{\text{kW}}{\text{kVA}}$$

Like a phasor, a power triangle could help us to determine appropriate values of capacitance to improve the power factor.

Example calculation

A 230 V 4 kW motor has a power factor rating of 0.4. Determine a suitably sized capacitor to improve the power factor to 0.85. Supply frequency is 50 Hz.

Let us first work out some values to see the extent of the problem.

If we ignore reactance and, therefore, power factor, this motor should draw a current of:

$$\frac{4000\,\text{W}}{230\,\text{V}} = 17.4\,\text{A}$$

But in reality it draws a current of:

$$\frac{4000\,\text{W}}{230\,\text{V} \times 0.4} = 43.5\,\text{A}$$

So, you can see that the reactance in the circuit causes the motor to draw 43.5 A instead of 17.4 A. This is a huge difference. By installing a correctly sized capacitor into the circuit, we could improve this, reducing the overall current demand. As the motor is an inductive load causing the current to lag, a capacitor will draw the current back towards unity and, therefore, closer to the value of true power.

To work out the size of capacitor needed, we need to determine the amount of reactive power the capacitor consumes. Remember, this reactive power drawn by the capacitor doesn't increase overall power demand, it simply off-sets the reactance caused by the impedance as it draws the current towards leading and, therefore, reducing reactance.

To determine the amount of reactive power, we need to draw a power triangle to show the relationship before correction.

Using a suitable scale, we draw true power to represent 4 kW. Then we measure an angle from this of $\cos^{-1} 0.4 = 66.4°$.

The line from this angle represents the apparent power. We then draw a line at 90° from the other end of the true power line. This line represents the reactive power. The two lines meet to form the power triangle, as seen in the diagram.

As we need to improve power factor to 0.85, we need to measure another angle from the true power line at $\cos^{-1} 0.85 = 31.79$ (32°). This line represents the new apparent power following correction, and gives a value by which reactance must be reduced.

By measurement, this line represents 6.6 kVA$_r$ so:

$$\frac{kVA_r \times 1000}{V} = I_C$$

so:

$$\frac{6.6 \times 1000}{230} = 28.70\,A$$

So, we need a capacitor that will theoretically draw a current of 28.70 A.

Then we need to carry out some of the previous calculations in reverse.

So:

$$\text{as } I_C = \frac{V}{X_C} \text{ then } X_C = \frac{V}{I_C} \text{ so} \frac{230}{28.70} = 8\,\Omega$$

And then:

$$X_C = \frac{1}{2\pi f C} \text{ so } C = \frac{1}{2\pi f X_C} = \frac{1}{2\pi \times 50 \times 8} = 398 \times 10^{-6}$$

A 398 µF capacitor is needed.

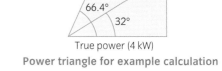

Power triangle for example calculation

Apparent power (10 kVA)
Reactive power (9.1 kVA$_r$)
66.4°
32°
True power (4 kW)

Value of power to be off-set by capacitor 6.6 kVA$_r$

Reactive power (9.1 kVA$_r$)

Apparent power (10 kVA)
66.4°
32°
True power (4 kW)

Power triangle showing correction of power factor

POWER FACTOR

The power factor is defined as the cosine of the angle by which the current leads or lags the voltage ($\cos \theta$). The power factor does not have a unit of measurement as it is a factor. The value can range from 0.01 to 0.99. A power factor of 1 is unity, the same as having a resistive circuit where the current rises and falls in phase with the voltage.

Power factors are used to express the effect of leading and lagging currents and many machines have a power-factor rating stated on the rating plate. The value is used to determine the current demand of the machine. As circuits with leading or lagging currents introduce

ACTIVITY

Use your calculator to find the cosine of 1°, 10°, 30°, 45°, 70° and 90°. Can you see that the greater the angle, the smaller the factor; the smaller the angle, the closer the factor is to 1, or unity?

If you reverse the process and choose a factor, you can determine the angle by using the $\cos^{-1}$ feature on your calculator. Try $\cos^{-1} 0.75$.

reactance, this creates the effect of additional loading and, therefore, additional current demand in a circuit. We need to understand and allow for this additional load when selecting equipment and cables for a circuit or installation.

Calculating power factor

To determine the values of the power factor at circuit level, we can apply the following equation:

$$\text{power factor} = \cos\theta = \frac{R}{Z}$$

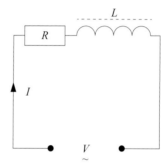

We will calculate power factor for this RL series circuit

The circuit shown in the diagram has the following values:

$R = 40\,\Omega$

$L = 38\text{ mH}$

Supply $= 230$ V 50 Hz

Example calculation

Determine:

a) the inductive reactance (X_L)

b) the impedance (Z)

c) the total circuit current (I)

d) the power factor and angle by which the current lags the voltage.

a) As $X_L = 2\pi fL$ then $X_L = 2 \times \pi \times 50 \times 38 \times 10^{-3} = 11.93\,\Omega$

b) As $Z = \sqrt{R^2 + X^2}$ then $Z = \sqrt{40^2 + 11.93^2} = 41.74\,\Omega$

c) As $I = \dfrac{V}{Z}$ then $I = \dfrac{230}{41.74} = 5.51$ A

d) As the power factor $= \cos\theta = \dfrac{R}{Z}$ then $\cos\theta = \dfrac{40}{41.74} = 0.958$ A

So if the power factor is 0.95, the angle by which the current lags the voltage (inductive circuit) is:

$$\cos^{-1} 0.958 = 16.7°$$

Power factors and impedance triangles

Power factors can also be determined from impedance triangles as the angle formed by the R and Z lines represents the angle by which the current leads or lags the voltage. The cosine of this angle is the power factor.

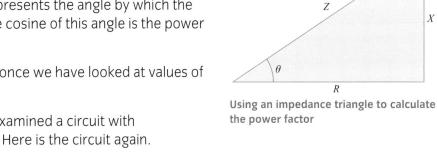

Using an impedance triangle to calculate the power factor

We shall look again at power factors once we have looked at values of power quantities in the next section.

In an earlier example (page 98), we examined a circuit with components in series and in parallel. Here is the circuit again.

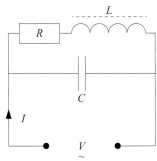

Circuit with components in series and in parallel

Recall the information given for this circuit:

$R = 12\ \Omega$

$L = 88$ mH

$C = 150$ μF

Supply = 230 V 50 Hz

As we have previously determined:

$$X_L = 2\pi fL$$

so:

$$X_L = 2\pi \times 50 \times 88 \times 10^{-3} = 27.64\,\Omega$$

and:

$$Z = \sqrt{R^2 + X_L^2}$$

so:

$$Z = \sqrt{12^2 + 27.64^2} = 30.13\,\Omega$$

The current in the inductive/resistive section of the circuit is:

$$I = \frac{V}{Z}$$

so:

$$I = \frac{230}{30.13} = 7.63\,\text{A}$$

The current drawn by the capacitor is based on the capacitive reactance:

$$X_C = \frac{1}{2\pi f C}$$

so:

$$X_C = \frac{1}{2\pi \times 50 \times 150 \times 10^{-6}} = 21.22\,\Omega$$

As:

$$I = \frac{V}{X_C}$$

then:

$$I = \frac{230}{21.22} = 10.83\,A$$

We can now go on to determine the power factor and angle of the lagging current in the series branch of the circuit in order to construct our phasor.

As the power factor $= \cos\theta = \dfrac{R}{Z}$

then $\cos\theta = \dfrac{12}{30.13} = 0.398$

So the angle is $\cos^{-1} 0.398 = 66.55°$ lagging (inductive circuit)

We can construct the phasor, firstly, by showing that the current drawn by the impedance (series branch) part of the circuit lags the voltage reference line by 66.55° to a scale value of 7.63 A.

We can then add the capacitor in parallel, drawing 10.83 A and leading the voltage by 90°. Forming a parallelogram from these two points gives us the point from which to draw the supply circuit current. This should work out to be 4.8 A to scale. From this, we can also see the angle at which it leads the voltage. The cosine of this angle is the power factor of the circuit.

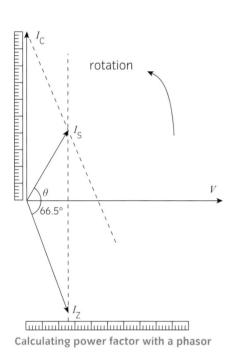

Calculating power factor with a phasor

ACTIVITY

Construct this phasor to scale and measure the new supply current angle. From this angle, determine the power factor of the circuit.

Assessment criteria

7.5 Explain what is meant by power factor correction and load balancing (neutral current)

POWER FACTOR CORRECTION

The effect of poor power factor absorbs part of the capacity of the generating plant and the distribution system, and as such distribution network operators charge industrial users on a kVA rather than a kW basis. It is therefore desirable for these users to keep the power factor of installed equipment as high as is practically possible.

Where the power factor is excessively low, power factor correction equipment can be installed which will produce economic savings and, where equipment and cables are operating at maximum rating, reduce the loading to safe limits and avoid overheating. Reduction of the current means system losses are reduced and improved voltage is usually obtained by the reduction in loading. Savings can also be made in cabling costs by improving the power factor.

In order to correct the power factor on the incoming supply it is necessary to measure it. The most convenient and easiest method of measuring power factor at any point in a circuit or installation is by means of a power factor indicator. If an indicator is not available an accurate calculation can be made by taking watt, volt and ampere measurement as shown in the diagram.

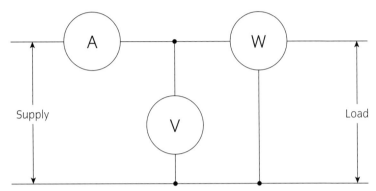

Connection of instruments to calculate PF

ACTIVITY

Power factor correction can be carried out by individual capacitors mounted at the load or by bulk correction at the intake. Name one advantage and one disadvantage of each arrangement.

The wattmeter is used to indicate true power (P in watts, W), whereas the volt and ammeter are used to calculate the apparent power (volt-amperes in VA).

Since:

$$PF = \frac{\text{true power}}{\text{apparent power}}$$

then the above connected meters will enable the PF to be calculated.

METHODS OF POWER FACTOR CORRECTION

Assessment criteria

7.6 Specify methods of power factor correction

The equipment available for improvement of the power factor can be divided into main two groups:

- rotary, eg synchronous motors (although this system is rarely used today)
- stationary, eg capacitors

Over-exciting a synchronous motor was often carried out in large factories during the postwar period to improve the overall power factor. Any large synchronous machines could be over-excited or run lightly to improve power factor. These motors would be used to supplement fixed banks of capacitors, which could be switched in to match the power factor throughout the day.

Correction by capacitors

The application of power factor correction capacitor banks is employed in electrical installations to correct the power factor.

This is due to the fact that most circuits are inductive in nature, which creates a lagging power factor. By adding power factor correction capacitors to the circuit, the kVA_r is reduced as the capacitive kVA_r cancels out the inductive kVA_r.

Power factor correction in many installations is achieved by a bank or banks of fixed capacitors. Alternatively, where the load changes or more accurate correction is required, power factor correction is achieved through automatically switched capacitors. Automatic switching units use monitoring technology to switch capacitors in and out of the load automatically as the load profile changes. These units are in banks, normally in multiples of $50kVA_r$, so that the first bank of fixed capacitors deals with the base load. Remember that going too far with capacitors causes a leading power factor, which is again chargeable.

Larger automatic switchable unit

Therefore, it is ideal to balance the load and switch in capacitors throughout the changing load profile to ensure the PF stays around 0.95. These units are normally installed at or near the intake position in a building in order to deal with the overall power factor correction.

As well as using capacitor banks to correct power factor at source, the power factor may be corrected through the equipment. Installing suitably rated capacitors in parallel with a load improves the power factor. As an example, fluorescent luminaires contain capacitors for this reason. The capacitor is connected between line and neutral in the luminaire. If the capacitor is removed, the luminaire will still operate but it will draw slightly more current due to the power factor.

NEUTRAL CURRENT IN THREE-PHASE AND NEUTRAL SUPPLIES

In a balanced three-phase system there is no requirement to have a star-point connection as the three phases have a cancellation effect on each other. Therefore the star point is naturally at zero current. While the load is balanced and the waveforms are symmetrical (not containing harmonics or other waveform distorting influences), this statement is relatively accurate. However, in practice, this is sometimes not the case.

Where the load is not in balance, different currents circulate in the load through the source winding and back. This gives rise to a change in star-point voltage, which can result in the system 'floating' away from its earthed reference point. In essence, a current will flow in the neutral. The three ways in which this current value could be determined are:

- by phasor, an accurate method that indicates the angle at which the maximum current occurs
- by calculation, which gives an accurate value
- by equilateral triangle, which gives a good indication of the value.

Determining current value by phasor

Assuming that a load has the current values $L_1 = 85$ A, $L_2 = 50$ A and $L_3 = 60$ A per phase, the current value can be determined by the following steps.

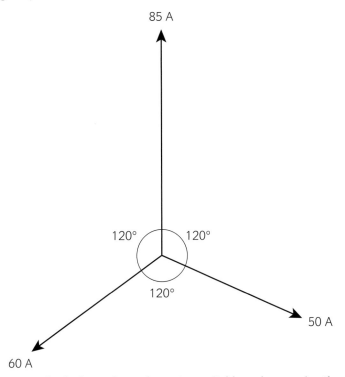

85 A

120° 120°

120°

50 A

60 A

Step 1 Construct a basic three-phase phasor to a suitable scale, ensuring the phases are 120° apart

Assessment criteria

7.7 Determine the neutral current in a three-phase and neutral supply

ASSESSMENT GUIDANCE

A three-phase heater or motor would provide a balanced load. Three identical houses connected to separate single-phase supplies probably wouldn't as each would have a different load.

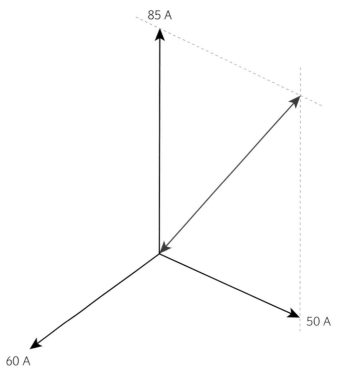

Step 2 Construct a parallelogram with two of the phases and draw a line from the centre point of the phasor to the point where the two new lines meet

ACTIVITY

Three equal 40 A loads are connected in star to a 400 V supply. The neutral current is zero. What will be the neutral current if:

a) one phase is disconnected

b) two phases are disconnected?

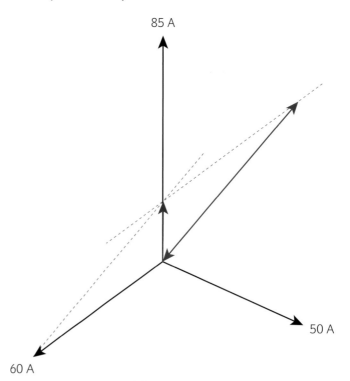

Step 3 Now construct another parallelogram between the new line and the remaining phase. Where the two new parallelogram lines intersect represents the neutral current value to the scale selected

Determining current value by equilateral triangle

The value of the neutral current can be determined using a scale drawing based on an equilateral triangle. If all phases are balanced, and therefore equal, all sides of the triangle are equal in length and meet to give equal angles of 60°.

If the phases are not balanced, there will be a gap at the top of the two sloping sides, which represents the neutral current. The triangle here represents a balanced system.

> **KEY POINT**
>
> Remember, an equilateral triangle is formed by three equal sides and has three equal angles of 60°.

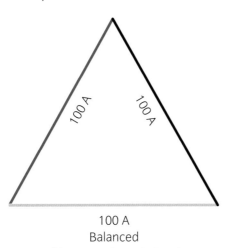

100 A
Balanced

A balanced system represented by an equilateral triangle

Now consider an unbalanced system with these values:

- L1 = 70 A
- L2 = 100 A
- L3 = 60 A

Draw the diagram, using an appropriate scale to represent the three current values.

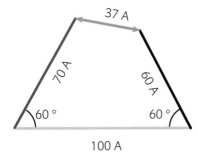

100 A

In an unbalanced system, the gap (shown in light blue) represents the neutral current

> **KEY POINT**
>
> When constructing scaled diagrams, it is crucial that you use a good scale and measure accurately.

The neutral current is represented by the gap left where the two sides do not meet (shown in light blue). In this example, that gap represents a current of approximately 37 A.

VOLTAGE AND CURRENT IN STAR AND DELTA CONNECTED SYSTEMS

Star (Y) and delta (Δ) configurations are used throughout the building services industry. Each configuration has different characteristics in terms of voltage and current values. Calculation of these values is essential in modern electrical engineering.

Voltage and current in star-connected systems

In a star (Y) connected load:

- the line current (I_L) flows through the cable supplying each load
- the phase current (I_P) is the current flowing through each load.

So:

$$I_L = I_P$$

and:

- the voltage between any line conductors is the line voltage (V_L)
- the voltage across any one load is the phase voltage (V_P)

so:

$$V_P = \frac{V_L}{\sqrt{3}} \text{ or } V_L = V_P \times \sqrt{3}$$

In a balanced three-phase system there is no need to have a star-point connection to neutral as the current drawn by any one phase is taken out equally by the other two. Therefore the star point is naturally at zero.

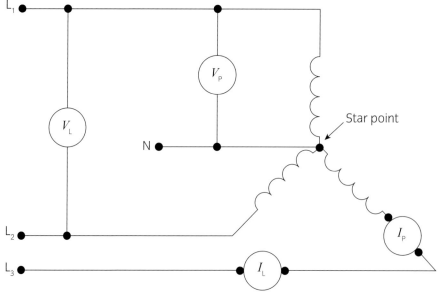

Star-connected load

So if a line current is 10 A, the phase current will also be 10 A. If the line voltage was 400 V, the phase voltage would be:

$$\frac{400}{\sqrt{3}} = 230\,\text{V}$$

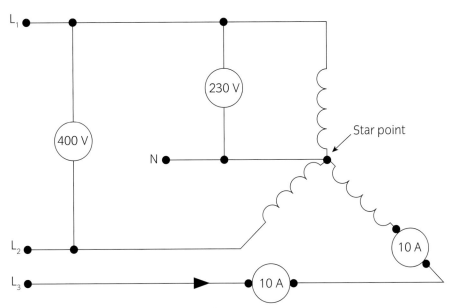

Common phase and line currents in load

Voltage and current in delta-connected systems

In a delta (Δ) connected load:

- the line current (I_L) flows through the cable supplying each load
- the phase current (I_P) is the current flowing through each load.

So:

$$I_P = \frac{I_L}{\sqrt{3}} \text{ or } I_L = I_P \times \sqrt{3}$$

and:

- the voltage between any line conductors is the line voltage (V_L)
- the voltage across any one load is the phase voltage (V_P)

so:

$$V_L = V_P$$

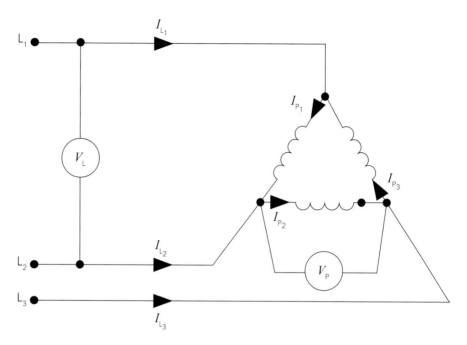

Delta-connected load similar to delta-connected supply

As there is no provision for a neutral connection, items such as delta motors would automatically be balanced – but complex loads on transmission systems could be unbalanced.

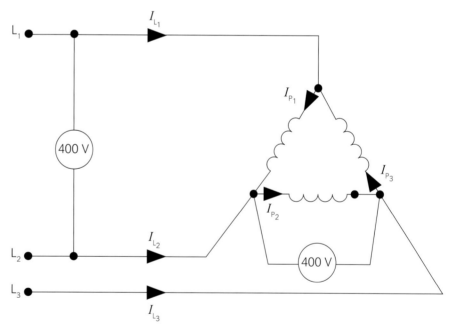

Phase and line currents in the load

Therefore if the phase current is 100 A and the phase voltage is 400 V, the line current can be calculated as follows:

$$I_{line} = 1.732 \times 100 = 173.2\,A$$

Understand the operating principles and applications of d.c. machines and a.c. motors

It is important to understand the differences in a.c. and d.c. machines and to appreciate where there are similarities in certain machine configurations.

HOW A.C. AND D.C. MACHINES OPERATE

Direct current (d.c.) machines were once the most popular type of machine because of the ability to control speed and direction. With advances in cheaper a.c. alternatives, d.c. machines were used less. Now that the parts and control devices for d.c. machines are cheaper, the use of d.c. is on the increase again. The competent electrician must therefore have a knowledge of d.c. machine operating principles.

There is no difference in the construction of d.c. motors and generators. They are rotating machines with three basic features: a magnetic-field system, a system of conductors and provision for relative movement between the field and the conductors.

The magnetic field in most d.c. machines is set up by the stationary part of the machine, called the field windings. The rotating part, known as the armature, is made up of multiple loops of cable linked to a commutator. Power is either delivered to (motor) or taken from (generator) the armature by brushes in contact with the moving commutator.

Assessment criteria

8.1 State the basic types, applications and describe the operating principles of d.c. machines

ACTIVITY

Losses occur in d.c. machines which reduce the efficiency. Name three such losses.

SmartScreen Unit 309
PowerPoint 7 and Handout 7

SmartScreen Unit 309
PowerPoint 8 and Handout 8

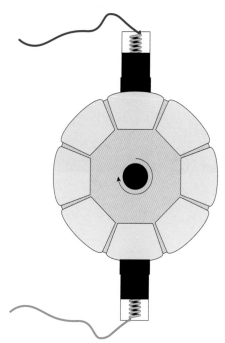

A segmented commutator with carbon brushes making contact. This commutator would have four loops wound around the armature.

d.c. generators

The d.c. generator is supplied with mechanical energy and gives out most of the energy, less losses, as electrical energy.

The d.c. generator has many loops and a multi-segmented commutator. With electricity flowing in the armature through brushes, the commutator reverses current flow as it passes from one pole to another so that the current in both conductors will always be the same.

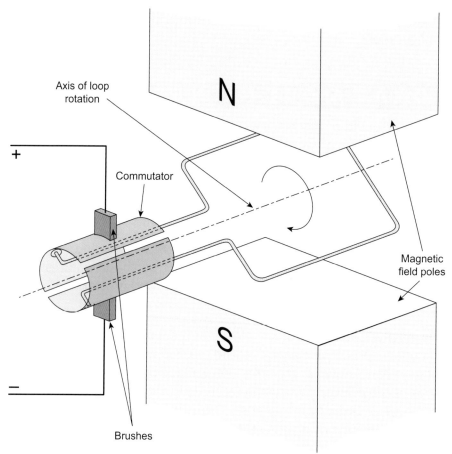

A simple generator arrangement

When the loops within the armature are rotated within the magnetic field, an emf is induced into the loop. The commutator ensures that the brushes are always in contact with the loop, which is in the strongest part of the magnetic field, at all times. This ensures a steady flow of direct current.

d.c. motors

The d.c. motor takes in electrical energy and provides mechanical power, less losses.

There are three types of d.c. motor: series, shunt and compound.

Series motors

Series motors are also known as universal motors as they can be used on alternating current as well. The field and armature in a series motor carry the same current and are capable of providing high starting torques. As the current is common to both parts, the windings are heavy gauge.

Series motors can be reversed in direction if a switch device is inserted between the field and armature, allowing simple reverse polarity of either the field or armature, but not both at the same time.

ASSESSMENT GUIDANCE

In the past, d.c. shunt generators were used in automobiles to keep the battery charged. In modern cars a three-phase alternator and bridge rectifier are used as this arrangement has a much higher output.

ASSESSMENT GUIDANCE

Series machines should always be connected to a load, otherwise they will run dangerously fast.

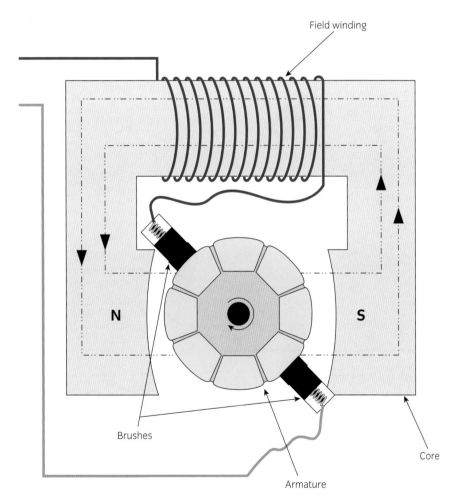

Simplified series motor arrangement

ASSESSMENT GUIDANCE

The torque of a series motor is proportional to the current squared.

$T \propto I^2$

Shunt motors

A shunt-connected d.c. motor consists of a field winding in parallel with the armature. This type of motor does not have the same common current characteristics as the series motor and therefore does not have a high starting torque. However, speed control of the shunt motor is considerably easier than the series motor as the field current can be controlled independently from the armature.

Shunt motors can be reversed in direction if the polarity of either the field or armature is reversed, but not both at the same time.

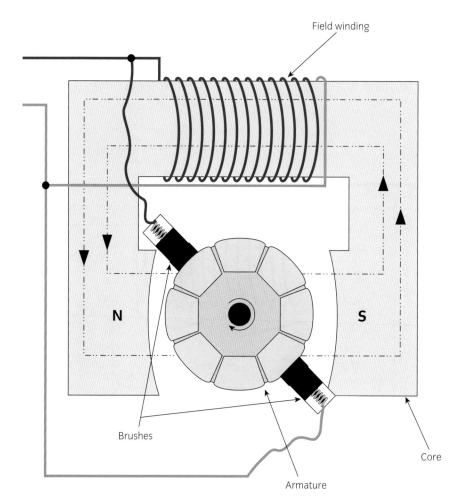

Simplified shunt motor arrangement

Compound motors

The compound motor is a mixture of the series and shunt motor circuits, offering the benefits of each type of machine, ie high starting torque and good speed control. To reverse a compound motor, the armature field must be reversed.

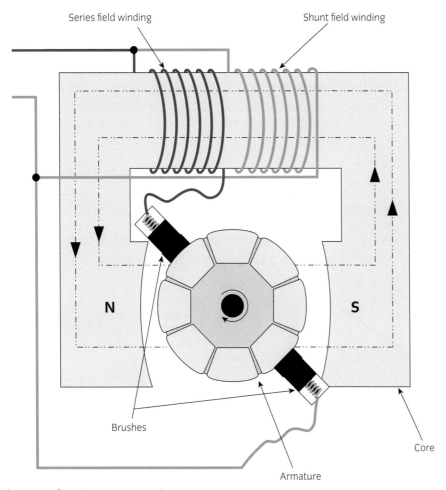

Compound motor arrangement

Applications of d.c. machines

Series-wound motors have excellent torque (load) characteristics and are used for applications such as dragline excavators, where the digging tool moves rapidly when unloaded but slowly when carrying a heavy load.

Shunt motors are best used where constant speed and torque are to be maintained, for example, on a production line, so that items placed on it do not affect the speed.

Compound motors offer the benefits of both series and shunt motors and have been used in older underground trains.

Direct current generators do not have many practical uses in their own right. However, they can provide a reliable energy supply directly into batteries or where a d.c. supply is required.

Assessment criteria

8.2 Describe the operating principles of:

- single-phase a.c. motors (capacitor start, induction start, universal)
- three-phase a.c. motors (squirrel cage, wound-rotor)
- inverter motors/variable frequency drives
- synchronous motors

HOW A.C. MACHINES OPERATE

The principle of operation for all a.c. motors relies on the interaction between a revolving magnetic field created in the stator by an a.c. current, with an opposing magnetic field either induced on the rotor or provided by a separate d.c. current source. This produces a torque that can be used to drive various loads.

The synchronous motor and induction motor are the most widely used types of a.c. motor. The difference between the two types of motor is that in a synchronous motor the rotation of the shaft is synchronised with the frequency of the supply current, and in an induction motor the rotor rotates slightly slower than the a.c. supply current in order to develop torque. The synchronous motor therefore operates at a precise speed.

How single-phase a.c. machines operate

A single-phase generator is composed of a single stator winding with one pair of terminals. With a single pair of rotating poles, the output waveform is as shown below.

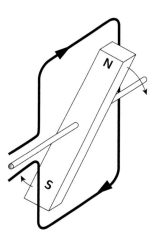
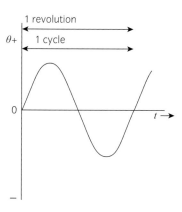

Two-pole machine and output

When a four-pole machine is used, the output waveform is changed as follows.

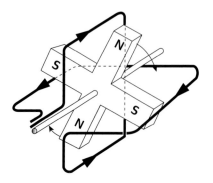
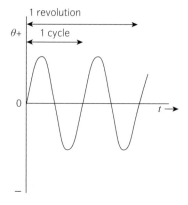

Four-pole machine and output

How three-phase a.c. machines operate

Three-phase a.c. motors have a number of advantages over their single-phase equivalents, including:

- smaller physical size for a given output
- steady torque output
- the ability to self-start without additional equipment.

The induction motor is the simplest and most common form of motor. The stator consists of a laminated body with slots for the field windings to pass through. The rotor is a laminated cylinder with conducting bars just below the surface.

In its simplest form, the rotor consists of a number of conductors passing through holes in a rotor drum. The ends are brazed together, causing the formation of a cage, which is why it is called a cage rotor.

Cage induction motors are cheaper and smaller, but produce less torque, than wound rotor induction motors. Wound rotor motors provide speed control via resistors and slip rings, but this is an inefficient method of controlling speed.

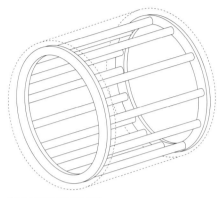

Cage arrangement

How cage induction motors work

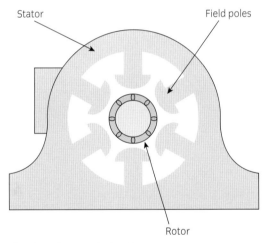

Stator

Field poles

Rotor

Component parts of a six-pole cage induction motor

If a rotor is placed in the field set up by the field windings in the stator, it will be cut by alternate north and south fields as the field rotates through one cycle. This creates an emf in the rotor, which causes current to flow in the conductor bars of the rotor. This current flow sets up a magnetic field, which causes the rotor to move and follow the rotating magnetic field in the stator.

The rotor rotates faster, heading towards **synchronous speed**. However, as the rotor speed increases, the difference between **rotor speed** and stator field speed reduces, causing the emf induced to reduce. The reduction in torque reduces the acceleration/velocity, meaning that synchronous speed cannot be met because no field cuts the rotor and, in turn, no emf or current in the rotor is induced.

ASSESSMENT GUIDANCE

The three-phase coils set up a rotating magnetic field in the stator. This can be reversed by changing over any two supply phases.

Synchronous speed

The speed at which the rotating field rotates around the field poles

Rotor speed

The actual speed at which the rotor rotates in revolutions/second (r/s)

Induction motors

The induction motor is the simplest form of alternating current (a.c.) motor. It is also known as the asynchronous motor.

Slip

The difference between the synchronous speed and the rotor speed expressed as a percentage or per unit value

Consider that a particular pole is north and that north then moves to the next pole (and the next) until there has been one complete rotation. The synchronous speed is the number of rotations completed in one second. The synchronous speed is affected by the supply frequency and the number of pairs of poles. For example, if a motor had six poles (three pairs) and the supply frequency was 50 Hz, the synchronous speed would be:

$$\text{synchronous speed } (n_s) = \frac{f}{p} = \frac{50}{3} = 16.67 \text{ r/s}$$

ACTIVITY

A three-phase a.c. alternator has four poles per phase. Calculate the speed in revs/sec needed to produce an output of:

a) 40 Hz
b) 50 Hz
c) 60 Hz.

This means that **induction motors** reach an ideal balancing velocity where there is sufficient **slip** to ensure that an emf is generated, resulting in a torque to turn the rotor. The fundamental operating principle of induction motors is that there has to be slip for the motor to work.

Slip may be represented as a percentage or a factor. If a motor with a synchronous speed of 16.67 revolutions/second had a slip of 8%, the rotor speed would be:

$$\text{rotor speed } (n_s) = \text{synchronous speed} - \left(\frac{\text{synchronous} \times \text{slip percentage}}{100} \right)$$

$$= 16.67 - \left(\frac{16.67 \times 8}{100} \right)$$

$$= 15.34 \text{ r/s}$$

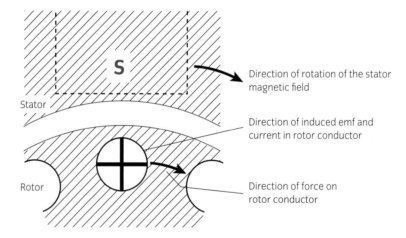

Induction motor principles

The speed of an induction motor can be varied by switching field pole pairs in and out.

Example

Calculate the synchronous speed of a two-, a four- and a six-pole motor fed from a 50 Hz supply.

For a two-pole motor = one-pole pair:

$$n_s = \frac{f}{p} = \frac{50}{1} = 50 \text{ r/s}$$

For a four-pole motor = two-pole pair:

$$n_s = \frac{f}{p} = \frac{50}{2} = 25 \text{ r/s}$$

For a six-pole motor = three-pole pair:

$$n_s = \frac{f}{p} = \frac{50}{3} = 16.67 \text{ r/s}$$

It can be seen that the speed of a motor is varied by the number of poles. However, electronic controls such as inverter drives are more effective and allow the speed to be varied over a wide range, matching it to the load requirements.

Wound rotor induction motors

In wound rotor induction motors, the rotor windings are connected through slip rings to external resistances. Adjusting the resistance allows control of the speed/torque characteristic of the motor by controlling the flow of induced current in the rotor. These motors can be started with low inrush current by inserting high resistance into the rotor circuit; as the motor accelerates, the resistance can be decreased.

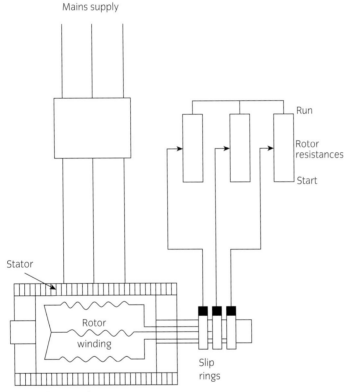

Wound rotor arrangement

The use of simple induction motors with readily available variable frequency drives means the wound rotor motor is less common, as changing the frequency will also affect the synchronous speed and therefore the rotor speed.

Where a process does not require a motor to run at full speed, a variable frequency drive (VFD) can be used to reduce the frequency and voltage to meet the requirements of the motor's load. Other names for a VFD are variable speed drive, adjustable speed drive, adjustable frequency drive, a.c. drive, microdrive and inverter.

Single-phase induction motors

As we have seen, three-phase motors use one phase to induce current into the rotor bars, with another phase, at a different polarity, creating the repulsion/attraction. However, the single-phase motor will not start by itself. This is due to the magnetic flux components being equal and opposite, cancelling out and leaving no torque to turn the rotor. Single-phase motors need to be modified to give the phase shift needed for the motor to start by itself.

To overcome the starting problem, single-phase motors have some form of additional start winding. The current in this winding can be made to lead or lag the main field winding by various methods. We saw on page 98 that a motor winding is an inductor and resistance. The phase angle can be changed by changing resistance values or by adding a capacitor.

The types of single-phase motors with phase-shift in a start winding are:

- split-phase induction motors
- capacitor-start motors
- shaded-pole motors.

Split-phase induction motors

The split-phase induction motor has a separate start winding connected to the main supply through a centrifugal switch.

Split-phase induction motor

The separate winding causes a slight phase shift by having a different resistance. This causes the rotation. The direction of rotation is determined by the polarity of the start winding, which is switched out by the centrifugal switch once a particular speed is achieved.

Capacitor-start motors

The capacitor-start motor is a variation of the split-phase induction motor. A starting capacitor is in series with the start winding, which creates a phase shift in the circuit due to the inductive and capacitive circuit formed by the winding and the capacitor.

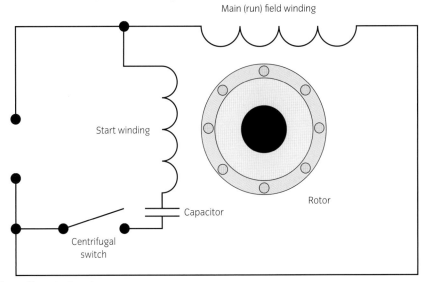

Main (run) field winding

Start winding

Capacitor

Centrifugal switch

Rotor

Capacitor start motor

In most capacitor-start motors, the capacitor is switched out of circuit by a centrifugal switch. The capacitor-start-and-run motor is a variant that does not switch out the capacitor.

Shaded-pole motors

The shaded-pole motor is a quite commonly used in devices, such as domestic appliances, that require low starting torque.

The motor is constructed with small single-turn copper shading coils, which create the moving magnetic field required to start a single-phase motor. A small section of each pole is encircled by a copper coil or strap; the induced current in the strap opposes the change of flux through the coil. This causes a time lag in the flux passing through the shading coil, creating an opposing pole to the main part of the pole. This effect is the same as having a phase shift causing an opposing polarity.

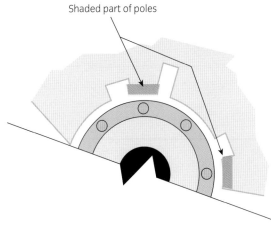

Shaded part of poles

Shaded-pole motor

Universal motors

An a.c. universal motor is very similar in construction to a d.c. series-wound motor. These devices combine the advantages of a.c. machines with some of the characteristics of d.c. As both field and armature currents reverse at the same time, it will work on an a.c. supply.

These motors have a high starting torque but, like the d.c. machine, tend to run very fast on no-load. In most cases a fan is fitted giving air resistance (windage) which limits the speed to a safe value.

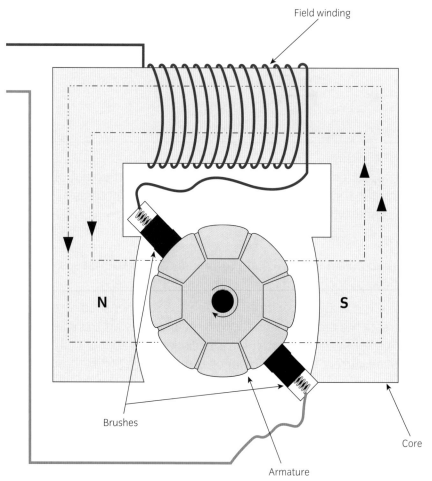

Section through a universal motor

APPLICATIONS OF A.C. MOTORS

Applications of single-phase a.c. machines

The commonly used shaded-pole motor is used where a high starting torque is not required, for example, in electric fans or drain pumps for washing machines and dishwashers, and in other small household appliances.

Universal motors are commonly used in small household appliances. such as food blenders, or power tools, such as drills, where smaller motors are beneficial.

Capacitor-start motors are commonly found in applications such as central-heating circulation pumps.

Inductive start motors are better suited to belt-drive applications due to poorer starting torque.

Applications of three-phase a.c. machines

a.c. generators

The a.c. generator or alternator is very widely used, eg in hybrid electrical vehicle drives, small and large-scale power generators, wind turbines or micro-hydro systems (small hydro-electric plant using stream water).

Induction motors

The cage induction motor is a simple, cost-effective induction motor. Advances in technologies for drive systems (eg thyristor (silicon-controlled rectifier)) and speed-control drives have enabled simple induction motors to replace more expensive wound rotor induction and some d.c. motors. Such technologies offer simple speed-controlled drives and reduced current starting for 'soft start' systems.

As a result, induction motors are used in a wide range of applications such as pumps, hoists, lifts and many other machines. The cage rotor is particularly useful because it has fewer parts that are subject to wear, having no brushes or slip rings.

Assessment criteria

8.3 State the basic types, applications and limitations of:

- single-phase a.c. motors (capacitor start, induction start, universal)
- three-phase a.c. motors (squirrel cage; wound-rotor)
- inverter motors/variable frequency drives
- synchronous motors

ASSESSMENT GUIDANCE

Induction motors are the work horses of industry. They are used in all kinds of machine drives.

Assessment criteria

8.4 Describe the operating principles, limitations and applications of motor control

HOW MOTOR STARTERS OPERATE

Motors are started in various ways. The choice of motor starting/control device depends on:

- available supply (single- or three-phase)
- motor start-up current
- starting speed.

Direct online starters (DOL)

These motor control devices literally switch a motor on or off with no varying speed or reduction in starting current. They are suitable for small, low-powered motors and can be used for single- or three-phase applications. The unit contains an electromagnetic coil that operates a contactor. The unit also incorporates overload contacts, finely tuned to the motor's start and running current to trip the device should an overcurrent occur. The coil that operates the contactor also provides undervoltage protection, as the contactor will open should a voltage lower than the coil voltage rating occur, eg during loss of supply. This means that the machine will not automatically restart unexpectedly should power resume.

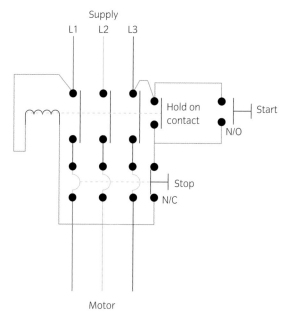

Direct online (DOL) starter

Star-delta starters

Star-delta starters must be connected to three-phase motors, which have six connection points at the motor. The motor is started by being put into a star connection, which reduces the starting current. Following a user-set time, which allows the motor to reach a particular speed, a time switch switches over the contactors automatically, putting the motor into delta connection and allowing full load current. The purpose of this is to reduce high current on start-up.

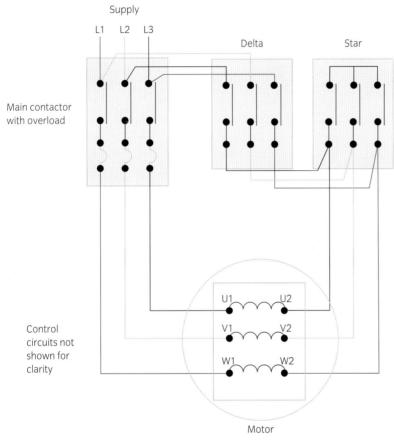

Star delta starter arrangement

ACTIVITY

What fraction of power is available in a star connection compared with a delta connection?

ASSESSMENT GUIDANCE

Star-delta starting requires access to all six ends of the motor windings.

To avoid short circuits, the star and delta contactors are normally physically interlocked by an electrical interlock and a mechanical connecting rod to prevent both contactors being in circuit at the same time.

Rotor resistance starters

This type of starting device works by introducing variable resistance to the rotor windings. It requires the motor to be a wound rotor induction motor, not a cage induction motor. The rotor windings are connected to an external variable resistance unit by slip rings and brushes. Variations in resistance can reduce start-up currents in the rotor.

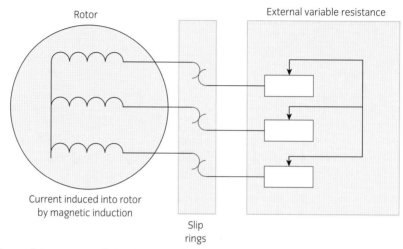

Rotor resistance control circuit (stator field winding and supply circuits are not shown for clarity)

Electronic motor starters

These motor soft-starters are used with a.c. motors to reduce the load and torque in the motor circuit during start-up. This is normally achieved by reducing the voltage; then, as the motor starts to achieve running speed, the voltage is ramped up. The term 'soft-start' also applies to the mechanical stresses placed on the components as they are also not subjected to intense starting forces.

Variable frequency drives (VFD) control the motor speed and torque of a.c. induction motors, by adjusting the frequency and voltage. They save energy and money by adjusting constant-speed devices such as pumps and fans to match the appropriate outputs.

Operating principles of motor control

Some motors use reverse-current braking and, where such reversal might result in danger, measures should be taken to prevent continued reversal after the driven parts come to a standstill at the end of the braking period. Further, where safety is dependent on the motor operating in the correct direction, means should be provided to prevent reverse operation. Where motor control systems incorporate overload devices, these devices should be checked to ensure the settings are correct and re-sets are correctly set to manual or automatic.

In this section you will find out about many types of electrical equipment in everyday use and how they work.

THE OPERATING PRINCIPLE OF ELECTRICAL EQUIPMENT

Electricity is an everyday commodity that most users give little thought to. The range and usage of electricity grows every day. New products continue to enter the market, and are added to the wide range of electrically powered equipment that people use in everyday life and very quickly taken for granted.

Solenoids

The strength of a magnetic field is proportional to the current flowing through it. Even with a high current passing through a conductor the field produced is relatively weak, in terms of useful magnetism. To obtain a stronger magnetic field a number of conductors can be added by turning or winding the cable.

The most common form of this arrangement is the solenoid, which consists of one long insulated conductor wound to form a coil. The winding of the coil causes the magnetic fields to merge into a stronger field similar to that of a permanent bar magnet. The strength of the field depends on the current and the number of turns.

Assessment criteria

9.1 Specify the main types and operating principles of the following electrical components:

- contactors
- relays
- solenoids
- over-current protection devices:
  - ☐ fuses (HRC, cartridge and re-wireable)
  - ☐ circuit-breakers
- RCDs
- RCBOs

SmartScreen Unit 309
PowerPoint 9 and Handout 9

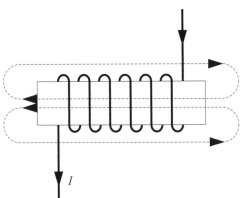

A cable wound around a tube: the current at the top moves away from the viewer and the current at the bottom moves towards the viewer

Placing a hinged, ferrous material attached to a spring in proximity to the magnetic field will cause that material to be drawn into the field when a very small current is applied to the solenoid.

This mechanical movement could be put to all sorts of uses, including:

- alarm or door bells
- relays
- contactors
- door hold/release systems
- fan shutter open/close drives
- circuit breakers
- residual current devices (RCD).

Some of these applications are described here.

Relays

A relay is an electrically operated switch, which uses an electromagnet to operate a set of contacts mechanically. This mechanical movement allows complete isolation of the switching from the initial signalling.

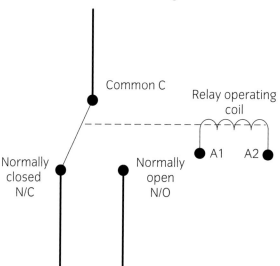

Relay showing contact positions

Relays are used often to control a circuit by a low-power signal in complete isolation from the larger circuit, or multiple circuits, being controlled. The first relays were used in long-distance telegraph circuits, repeating the signal coming in from one circuit and re-transmitting it to another. Relays were then used extensively in telephone exchanges to perform logic functions and operations.

With modern technological advances, not all relays consist of a coil operating a set of magnets. Solid-state relays either replace or are available in conjunction with electromechanical relays. Solid-state relays control electrical circuits despite having no moving parts. Instead, they use a semiconductor device to perform the switching.

ASSESSMENT GUIDANCE

A typical application of a contactor is to control a large heating load by the use of a thermostat. The thermostat can operate at extra low voltage but still control a large heating load.

Contactors

A relay that can handle the high power used to control directly an electric motor or other loads is called a contactor. There is little difference between a relay and a contactor, but generally contactors are devices that switch heavier loads on and off, whereas relays either switch or divert (like a two-way switch) lower current loads.

PROTECTIVE DEVICES

Protective devices may be one or a combination of:

- fuses
- circuit breakers (CBs)
- residual current devices (RCDs).

Fuses

Fuses have been a tried and tested method of circuit protection for many years. A fuse is a very basic protection device that melts and breaks the circuit should the current exceed the rating of the fuse. Once the fuse has 'blown' (ie the element in the fuse has melted or ruptured), it needs to be replaced.

Fuses have several ratings.

- I_n is the nominal current rating. This is the current that the fuse can carry, without disconnection, and without reducing the expected life of the fuse.
- I_a is the disconnection current rating. This is the value of current that will cause the disconnection of the fuse in a given time.
- Breaking capacity (kA) rating. This is the current up to which the fuse can safely disconnect fault currents. Any fault current above this rating may cause the fuse and carrier to explode.

BS 3036 rewirable fuses

In older equipment, the fuse may be just a length of appropriate fuse wire fixed between two terminals. There are increasingly fewer of these devices around as electrical installations are rewired or updated.

One of the main problems associated with rewirable fuses is the overall lack of protection, including insufficient breaking capacity ratings. Another major problem is that the incorrect rating of wire can easily be inserted when replacing the fuse wire, leaving the circuit underprotected.

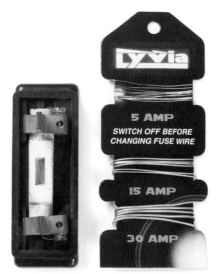

Rewirable fuse and fuse wire card showing how wrong wire can easily be used

BS 88 fuses

These modern fuses are generally incorporated into sealed cylindrical ceramic bodies (or cartridges). If the element inside blows, the whole cartridge needs to be replaced. Although these devices have fixed time current curves, they can be configured to assist discrimination. The benefit of BS 88 and similar fuses is their simplicity and reliability, coupled with high short-circuit breaking capacity.

Within some types of BS 88 fuse, usually the bolted type, there may be more than one element. The purpose of this is to minimise the energy from a single explosion, should the fuse be subjected to high fault currents. Instead there will be several smaller explosions, allowing these devices to handle much higher fault current (up to 80 kA).

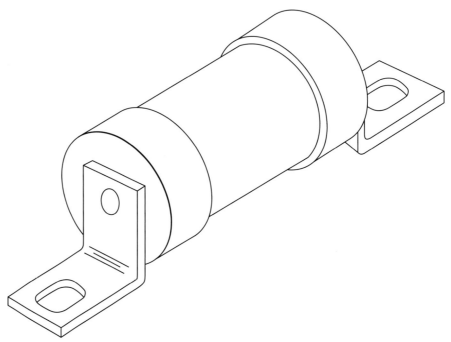

BS 88 bolted type fuse

Other BS 88 devices may be the clipped type, which do not have the two bolt tags. They are simply barrel shaped and slot into place in the carrier. They are often called cartridge fuses.

Another type of cartridge fuse is the BS 1362 plug fuse. These are fitted into 13 A plugs and are available in a range of ratings. Typical ratings are 3 A, 5 A and 13 A.

Circuit breakers

Circuit breakers (CB) have several ratings.

- I_n is the nominal current rating. This is the current that the device can carry, without disconnection and without reducing the expected life of the device.

- I_a is the disconnection current. This is the value of current that will cause the disconnection of the device in a given time.
- I_{cn} is the value of fault current above which there is a danger of the device exploding or, worse, welding the contacts together.
- I_{cs} is the value of fault current that the device can handle and remain serviceable.

Circuit breakers are thermomagnetic devices capable of making, carrying and interrupting currents under normal and abnormal conditions. They fall into two categories: miniature circuit breakers (MCBs), which are common in most installations for the protection of final circuits, and moulded-case circuit breakers (MCCBs), which are normally used for larger distribution circuits.

Both types work on the same principle. They have a magnetic trip and an overload trip, which is usually a bimetallic strip. If a CB is subjected to overload current, the bimetallic strip bends due to the heating effect of the overcurrent. The bent strip eventually trips the switch, although this can take considerable time, depending on the level of overload.

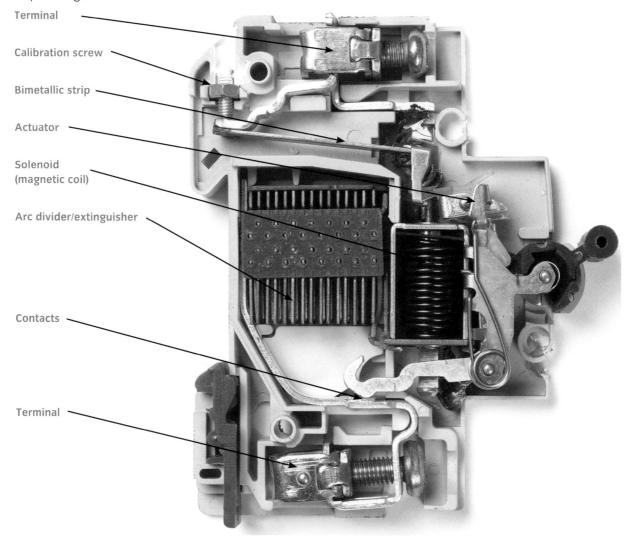

Terminal

Calibration screw

Bimetallic strip

Actuator

Solenoid (magnetic coil)

Arc divider/extinguisher

Contacts

Terminal

Section through a circuit breaker

Miniature circuit breakers (MCBs)

These thermomagnetic devices have different characteristics, depending on their manufacture. They generally have a lower prospective short-circuit current rating than a high-rupturing capacity (HRC) fuse, ranging from approximately 6 kA to 10 kA. Specialist units are available for higher values.

The operating characteristics of MCBs can be shown in graphical form by a time–current curve. MCBs are shown to be generally faster acting than the standard BS 88 fuses. A CB has a curve, then a straight line, whereas the BS 88 fuse is fully curved. This demonstrates the two tripping mechanisms in a CB. The magnetic trip is represented by the straight line on the graph, indicating that a predetermined value of fault current will disconnect the device rapidly. The curve represents the device's thermal mechanism. Like a fuse, the thermal mechanism reacts within a time specific to the overload current. The bigger the overload, the faster the reaction.

▼ **Table 7.2.7(i)** Rated short-circuit capacities

Device type	Device designation	Rated short-circuit capacity (kA)	
Semi-enclosed fuse to BS 3036 with category of duty	S1A S2A S4A	1 2 4	
Cartridge fuse to BS 1361 type I type II		16.5 33.0	
General purpose fuse to BS 88-2		50 at 415 V	
BS 88-3 type I type II		16 31.5	
General purpose fuse to BS 88-6		16.5 at 240 V 80 at 415 V	
Circuit-breakers to BS 3871 (replaced by BS EN 60898)	M1 M1.5 M3 M4.5 M6 M9	1 1.5 3 4.5 6 9	
Circuit-breakers to BS EN 60898* and RCBOs to BS EN 61009		I_{cn} 1.5 3.0 6 10 15 20 25	I_{cs} (1.5) (3.0) (6.0) (7.5) (7.5) (10.0) (12.5)

* Two short-circuit capacities are defined in BS EN 60898 and BS EN 61009:

I_{cn} the rated short-circuit capacity (marked on the device).
I_{cs} the in-service short-circuit capacity.

Rated short circuit capacities of protective devices (from the On-Site Guide, IET)

Moulded-case circuit breakers (MCCBs)

Although moulded case circuit breakers (MCCBs) work on the same principle as MCBs, the moulded case construction and physical size of MCCBs gives them much higher breaking capacity ratings than those of MCBs. Many MCCBs have adjustable current settings.

Residual current devices (RCDs) and residual current circuit breakers with overload (RCBOs)

Residual current devices (RCDs) operate by monitoring the current in both the line and neutral conductors of a circuit. If the circuit is healthy with no earth faults, the toroidal core inside the device remains balanced with no magnetic flux flow. If a residual earth fault occurs in the circuit, slightly more current flows in the line conductor compared to the neutral. If this imbalance exceeds the residual current setting of the device, the flux flowing in the core is sensed by the sensing coil, which induces a current to a solenoid, tripping the device.

ACTIVITY

What rating of RCD would be used to provide additional protection?

KEY POINT

Remember that RCDs only offer earth fault protection not short-circuit (line to neutral) or overload protection. If an RCD is fitted to a circuit, appropriate protective devices must also be installed to offer short-circuit and overload protection.

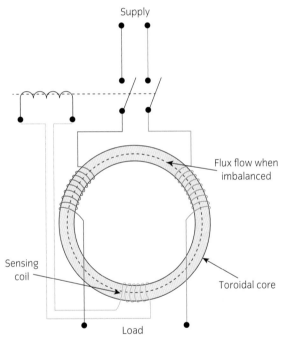

Internal circuit diagram for an RCD

Residual current breakers with overload (RCBOs) combine an overcurrent protective device with a RCD in the body of the CB.

Unlike CBs, RCDs and RCBOs have a test button, which should be pressed at very regular intervals to keep the mechanical parts working effectively. If the mechanical components in a CB stick, there is not much concern as the energy needed to trip a CB is large enough to unstick any seized parts. As RCDs and RCBOs operate under earth fault conditions, with relatively small residual currents, there may not be enough energy to free any seized parts.

RCBO to BS EN 61009

Assessment criteria

9.2 Describe how the following components are applied in electrical systems/equipment and state their limitations:

- contactors
- relays
- solenoids
- over-current protection devices:
  - fuses (HRC, cartridge and re-wireable)
  - circuit-breakers
- RCDs
- RCBOs

APPLICATION OF ELECTRICAL COMPONENTS IN ELECTRICAL SYSTEMS

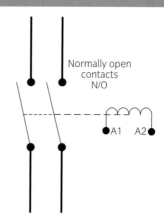

Typical contactor

Solenoids

Solenoids have a number of functions. They are often used as electrical–mechanical transducers (converters), ie they convert an electrical signal into mechanical action. This can be as some form of limit switch that trips a non-automatically resettable device or, more commonly, to operate a control valve on heating and other similar systems.

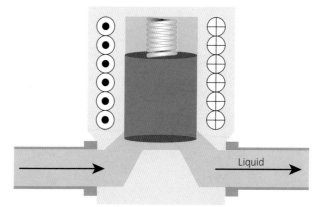

Solenoid valve

Solenoids are also often used in electromagnetic locking devices, either to engage or to retract the locking mechanism. Where safety is essential, such solenoids are usually positioned so that the system drops to a safe position if the electrical circuit fails. For example, a door release magnet will release the door in the event of a fire.

Relays

A relay is used so that one circuit, normally a low-current circuit, controls another by use of remote contacts. Some relays operate a large number of contacts, switching multiple circuits, with complete electrical isolation of the switching circuit from the operating circuit. Others switch high-current circuits using either low-power circuits or even extra low-voltage circuits, which are in turn controlled by logic devices such as programmable logic controllers (PLCs).

Contactors

The term 'contactor' is often used instead of 'relay'; however, 'contactor' is more accurately used for a large relay operating large loads, such as a motor.

In the case of motor starters such as the direct online (DOL) starter (see page 126), the contactor is also coupled to an overload device. The contactor itself provides control (on and off) as well as undervoltage protection, which is required where the loss of supply and subsequent restoration may cause danger. In the case of a motor or machine, the machine cannot restart after a loss of supply until someone physically pushes the start button on the starter.

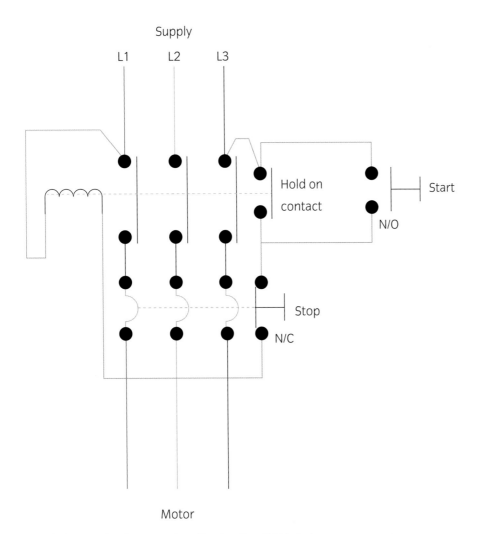

A typical example of a contactor: direct online (DOL) starter arrangement

BS 3036 rewirable fuses

Unlike most other protective devices, the BS 3036 fuse arrangement does not have a very accurate operating time or current as it is dependent upon factors such as age, level of oxidation on the element and how it has been installed (eg whether it was badly tightened, open to air movement).

A range of BS 3036 rewirable fuses: 5 A (white), 15 A (blue) and 30 A (red)

The lack of reliability of these fuses is a concern to designers and duty holders. Due to the lack of sensitivity, special factors have been applied to Appendix 4 of BS 7671 Requirements for Electrical Installations (the IET Wiring Regulations) to account for these fuses. This rating factor to be applied (C_f) is 0.725.

BS 88 fuses

High-rupturing capacity (HRC) or high-breaking capacity (HBC) fuses are common in many industrial installations. They are also very common in switch fuses or fused switches controlling specific items of equipment. They are particularly suited to installations with a high prospective fault current (I_{pf}) as they have breaking capacities of up to 80 kA.

BS 88 fuses come in two categories:

- gG for general circuit applications, where high inrush currents are not expected

- gM for motor-rated circuits or similar, where high inrush currents are expected.

ACTIVITY

Identify two other rewirable fuse carrier ratings and colours.

Miniature circuit breakers (MCBs)

There are three common types of MCB: Type B, Type C and Type D. The difference between the devices is the value of current (I_a) at which the magnetic part of the device trips. The different types are selected to suit loads where particular inrush currents are expected.

Type B trips between three and five times the rated current (3 to 5 × I_n). These MCBs are normally used for domestic circuits and commercial applications where there is no inrush current to cause it to trip. For example, the magnetic tripping current in a 32 A Type B CB could be 160 A. So $I_a = 5 × I_n$. These MCBs are used where maximum protection is required and therefore should be the choice for general socket-outlet applications.

Type C trips between five and ten times the rated current (5 to 10 × I_n). These MCBs are normally used for commercial applications where there are small to medium motors or fluorescent luminaires and where there is some inrush current that would cause the CB to trip. For example, the magnetic tripping current in a 32 A Type C CB could be 320 A. So $I_a = 10 × I_n$.

Type D trips between ten and twenty times the rated current (10 to 20 × I_n). These MCBs are for specific industrial applications where there are large inrushes of current for industrial motors, x-ray units, welding equipment, etc. For example, the magnetic trip in a 32 A Type D CB could be 640 A. So $I_a = 20 × I_n$.

ASSESSMENT GUIDANCE

In a fused switch the fuses are mounted on the moving contacts. In a switch fuse, the fuse and switch are in series and the fuse does not move.

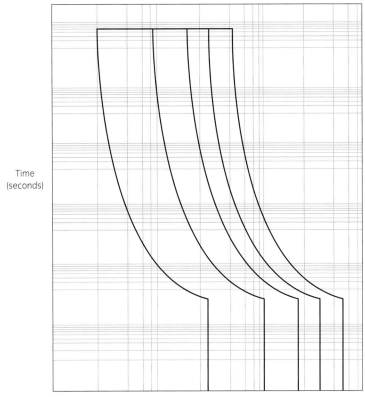

Time (seconds)

Prospective fault current (amperes)

Sample time–current characteristic graph which is found in Appendix 3 of BS 7671

Moulded-case circuit breakers (MCCBs)

MCCBs are available in various ranges. Lower-cost simpler versions are thermomagnetic with no adjustment. Other devices have electronic trip units and sensitivity settings or the ability to be de-rated.

Most MCCBs are used on larger circuits or distribution circuits where larger prospective short circuits are likely but the flexibility of an electronic trip is also required.

Residual current devices (RCDs) and residual current circuit breakers with overload (RCBOs)

RCDs are sensitive devices, typically operating on earth fault currents as low as 30 mA and with response times as fast as 20–40 ms.

The use of such a device is useful for supplies to portable electrical equipment where the risk of shock to the user might be higher than would otherwise be the case. BS 7671:2008 provides for wider use of RCDs, both for indoor circuits as well as for socket outlets intended to supply portable electrical equipment outdoors, ie outside the zone of earthed equipotential bonding.

Lighting is a specialist area in electrical installations work. Many designers and installers rely on specialists to manage the design of lighting, but there are some basic areas of knowledge that an electrician needs to know in order to install and maintain luminaires effectively.

LAWS OF ILLUMINATION AND ILLUMINATION QUANTITIES

Two laws explain how light behaves when it is emitted from a luminaire onto a surface: the inverse square law and the cosine law. First, look at the terms and units used to explain and quantify lighting.

SmartScreen Unit 309
PowerPoint 10 and Handout 10

Assessment criteria

10.1 Explain the basic principles of illumination and state the applications of:

- inverse square law
- cosine law
- lumen method

Lighting terms and units

Term	Symbol	Unit	Description
Luminous intensity	I	candela (cd)	The amount of light emitted per solid angle or in a given direction
Luminous flux	F	lumen (lm)	The total amount of light emitted from a source
Illuminance	E	lumens per metre^2 (lux)	The amount of light falling on a surface
Efficacy	K	lumens per watt (lm/W)	This is a term is used to measure the **efficiency** of a lamp or luminaire. It compares the amount of light emitted to the electrical power consumed.
Maintenance factor Or Light loss factor	Mf llf	none	These factors are used to de-rate the light output of a lamp, allowing for dust. The factor used depends on the environment. An average office environment would have a factor of 0.8 whereas a factory where lots of dust accumulates may be 0.4.
Coefficient of utilisation or utilisation factor	Uf	none	This factor takes the surfaces in a room, such as walls and ceilings, into account. Emitted light bounces off walls that reflect light well, making more effective use of the light. An average factor for a room is 0.6. The lighter the colour of the room, the higher the factor.
Space–height ratio			This ratio is used to determine how close together luminaires need to be, taking into account their height from a given surface, in order to illuminate a room with an even spread of light from multiple luminaires.

Efficiency

The ratio between power in and power out, measured in the same unit.

Efficacy

The ratio of power in and power out, measured in two different units. For example, the ratio of light output in lumens to the electrical power measured in watts.

Some lamps, such as light-emitting diode (LED) lamps, are rated in lumens whereas others are rated in candela. Take care with this as the choice of lamp depends on its application. To illuminate a particular point, such as a kitchen work surface, the candela rating is important as it rates the intensity of light in a particular direction. For illuminating a general area, such as a driveway, the luminous flux (lumens) is a better indicator as it measures the total light output in all directions.

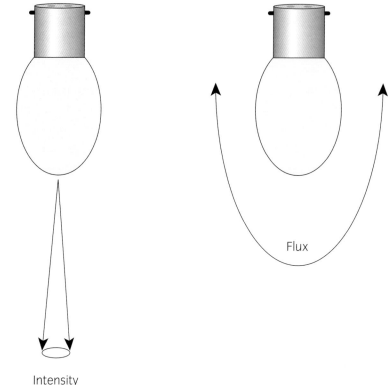

Flux

Intensity

Candela is the rating of intensity. The rating of total light output in all directions is the lumen.

ACTIVITY

A luminaire emits 1250 candela in all directions. Calculate the illuminance at a) 2.5 m and b) 5 m directly below the luminaire.

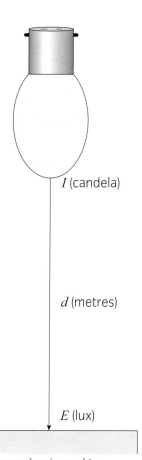

I (candela)

d (metres)

E (lux)

The inverse square law is used to determine the illuminance (lux) of a surface

Inverse square law

The amount of light falling onto a surface changes depending on the distance from the light source. If you hold a torch just above a surface, the light falling on the surface is intense. As you move the torch further away, the light directly below the torch becomes less intense because the light spreads. The amount of light on the surface is the illuminance. To determine the amount of light on the surface directly below the source, use the inverse square law.

$$E = \frac{I}{d^2} \text{ (lux)}$$

where:

I = luminous intensity, in candela (cd)

E = illuminance on the surface, in lux

d = distance between the lamp and surface, in metres (m).

Example

A 1000 cd light source is suspended above a level plane. Calculate the illuminance of the surface at 2 m and 4 m from the source.

At 2 m:

$$E = \frac{I}{d^2}$$

Therefore:

$$E = \frac{1000}{2^2} = 250 \text{ lux}$$

> **ASSESSMENT GUIDANCE**
>
> The inverse square law and cosine law are basically the same. It is just that the cosine of 0° is 1, so it has no effect on the calculation.

At 4 m:

$$E = \frac{I}{d^2}$$

Therefore:

$$E = \frac{1000}{4^2} = 62.5 \text{ lux}$$

Cosine law of illumination

When light falls obliquely on a surface, not at right angles to it, the light spreads over an increasing area as the angle (θ) between the perpendicular to the surface and the direction of the light, increases.

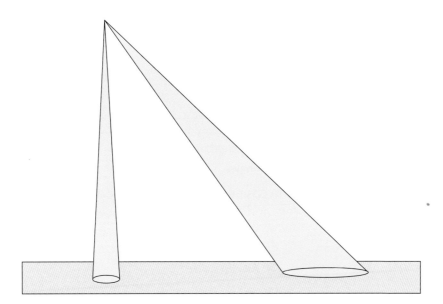

The spread of light at different angles

To calculate illumination in such cases, use the cosine law, which takes the additional area illuminated into account. It is expressed as:

$$E = \frac{I}{d^2} \times \cos\theta \text{ (lux)}$$

where:

I = luminous intensity, in candela (cd)

E = illuminance on the surface, in lux

d = distance between the lamp and surface, in metres (m)

$\cos\theta$ = cosine of the angle at which the light is emitted from the lamp.

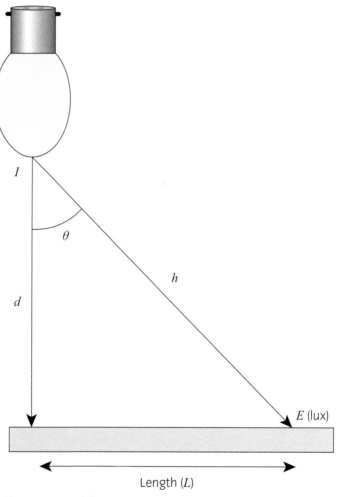

Calculating illumination at different angles

If the angle is unknown, the cosine of the angle can be determined by using Pythagoras' theorem and trigonometry.

As:

$$h = \sqrt{d^2 + L^2}$$

and:

$$\cos\theta = \frac{d}{h}$$

Example

A 1200 cd light source is suspended above a level plane. Calculate the illuminance of the surface at 2 m directly below the source (E_1) and then calculate the illuminance on a surface 4 m away (E_2).

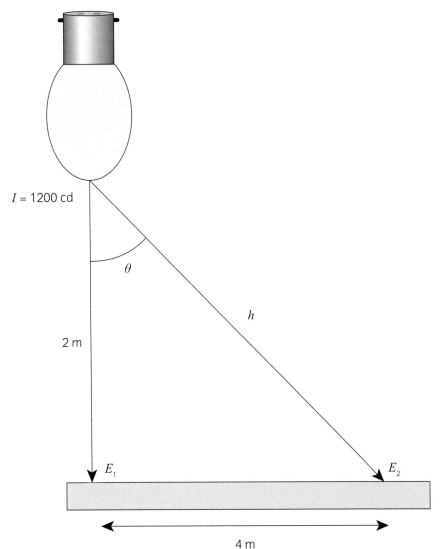

$I = 1200$ cd

θ

h

2 m

E_1

E_2

4 m

Calculating illuminance directly below the source and 4 m away

$$E_1 = \frac{I}{d^2}$$

Therefore:

$$E_1 = \frac{1200}{2^2} = 300 \text{ lux}$$

To determine E_2, find cos θ.

So:

$$h = \sqrt{d^2 + L^2}$$

Therefore:

$$h = \sqrt{2^2 + 4^2} = 4.47\,m$$

and

$$\cos\theta = \frac{d}{h}$$

Therefore:

$$\frac{2}{4.47} = 0.44$$

Therefore:

$$E_2 = \frac{1200}{4.47^2} \times 0.44 = 26.4\,\text{lux}$$

Lumen method

As can be seen from the cosine law, the level of light varies across an area due to the distance and angle of the light source. When a desired level of illuminance is specified, an average figure is used across the area or working surface.

Lighting guides give values of illuminance that are suitable for use in various areas. The average value is usually quoted in lux.

The lumen method is used to determine the number of lamps that should be installed for a given area or room to achieve a specific average illuminance level.

Guided average illuminance levels for different activities or areas

Activity or area	Illumination (lux, lumen/m²)
Public areas with dark surroundings	20–50
Working areas where visual tasks are only occasionally performed	100–150
Warehouses, homes, theatres, archives	150
Classrooms	250–350
Normal office work, computer work, study library	350–450
Supermarkets, mechanical workshops	750
Normal drawing work, detailed mechanical workshops	1000
Detailed drawing work, very detailed mechanical works	1500–2000
Performance of very prolonged and exacting visual tasks	5000–10 000

Calculating for the lumen method

The lumen method is appropriate for use in lighting design if the luminaires are to be mounted overhead in a regular pattern.

The luminous flux output (lumens) of each lamp needs to be known, as well as details of the luminaires and the room surfaces.

Usually, the **illuminance** will already have been specified by the designer, eg office 350–450 lux.

The formula is:

$$N = \frac{E_{average} \times A}{Mf \times Uf \times F}$$

where:

$E_{average}$ = average illuminance over the horizontal working plane, in lux

N = number of luminaires required

F = luminous flux of each luminaire selected, in lumens (as declared by the manufacturer)

Uf = utilisation factor based on the reflectance of the room walls, ceiling and work surface

Mf = maintenance or light loss factor (llf)

A = area to be illuminated.

Example

Determine the number of luminaires required in an office measuring 18 m by 20 m. The room is to be illuminated using recessed modular luminaires with a luminous flux given as 4800 lumens per fitting. The room has a white ceiling and walls painted a light colour; as a result, the utilisation factor is 0.9. As dust in an office is minimal, the maintenance factor is 0.8. The average illuminace desired is 400 lux.

Using:

$$N = \frac{E_{average} \times \text{area}}{Mf \times Uf \times F}$$

$$N = \frac{400 \times (18 \times 20)}{0.8 \times 0.9 \times 4800} = 41.66 \text{ or } 42 \text{ luminaires}$$

ASSESSMENT GUIDANCE

The actual number of fittings installed may be slightly higher than the calculated number to give equal numbers in each row.

OPERATION OF LUMINAIRES

Space-height ratio

The space–height ratio determines how far apart luminaires should be in relation to their intended mounting height, or visa versa. The ratio depends on the particular luminaire and is determined by the manufacturer.

Example

If the space–height ratio of a luminaire is 3:2 and the mounting height is 2.4 m above the working plane (the area where the maximum illuminance is needed), determine the distance needed between luminaires.

$$\frac{S_r}{H_r} = \frac{S}{H}$$

where:

S_r is the space ratio

H_r is the height ratio

S is the actual spacing between luminaires (centre to centre)

H is the height the luminaires are mounted.

$$\frac{3}{2} = \frac{S}{2.4}$$

So:

$$\frac{3 \times 2.4}{2} = S = 3.6\,m$$

APPLICATION OF LUMINAIRES

Efficacy

Lamps are given an efficacy rating based on the amount of luminous flux (in lumens) emitted by the lamp for every watt of power consumed by the lamp, including losses (in watts). An efficacy rating is a good indication of a particular lamp's energy efficiency. The higher the efficacy rating, the better the energy efficiency.

$$\text{efficacy}\ \frac{lm}{W} = \frac{\text{light output (lm)}}{\text{electrical input (W)}}\ lm/W$$

The purpose of a lamp is to produce light. Lamps also produce heat during operation. As energy is required to produce the heat, this counts as a loss of energy because heat is not the intended product of the lamp.

In order for buildings to comply with current Building Regulations or the Code for Sustainable Homes, guidelines are given for the minimum lumens of light per circuit watt consumed.

Example

The light output from a lamp is 8000 lumens and the input power is 150 watts. Calculate the efficacy of the lamp.

$$\text{efficacy} \frac{\text{lm}}{\text{W}} = \frac{8000}{150} = 53.33 \text{ lm/W}$$

Indication of the efficacy of different lamp types. Consult manufacturer's data for accurate ratings for particular lamps.

Lamp type	Efficacy (lm/W)
60 W GLS incandescent lamp	14
100 W tungsten halogen	16
50 W high-pressure sodium (SON)	120
18 W compact fluorescent (CF)	61
70 W metal halide	64
20 W T5 fluorescent tube	99
2 W LED	100
90 W low-pressure sodium (SOX)	180

Colour rendering

The term 'colour rendering' describes the ability of a lamp or luminaire to keep objects looking their true colour. Some lamps emit orange light and so objects lit by the lamp appear orange.

Colour rendering is very important when selecting luminaires, depending on their application. For example, general amenity lighting or street lighting does not require good colour rendering unless closed circuit TV (CCTV) is present, in which case this factor is important. Studies have shown that better colour rendering of street lighting in some areas can reduce crime, as criminals are aware that they can be identified more easily.

ACTIVITY

Using the internet, look up the current requirements for lighting efficacy according to the Code for Sustainable Homes.

KEY POINT

Just because a lamp has a good efficacy rating does not necessarily mean it is suitable. For example, the SOX lamp has extremely good efficacy but very poor colour rendering as everything illuminated by it looks orange!

ASSESSMENT GUIDANCE

A low-pressure sodium lamp over a dining table would produce a high level of illumination but the food would look terrible.

Colour rendering is also important in shops. Installing the correct type of fluorescent tube can make clothes look vibrant, food look appetising and also reduce eye strain. There are many types of fluorescent tube colours available, from warm white to daylight and colouright tubes.

More information relating to the application of different types of luminaires and lamps is given below.

OPERATION OF LAMPS

There are several types of lamp, each of which works in different ways:

- incandescent lamps
- discharge lamps
- compact fluorescent lamps
- LED lamps.

ASSESSMENT GUIDANCE

General lighting service (GLS) lamps have largely been replaced by compact fluorescent lamps, due to the low efficacy of the GLS types.

Incandescent lamps

This is the simplest form of lamp, with a current passed through a filament. The filament gets white hot and therefore emits light.

Tungsten filament lamps

Tungsten has a high melting temperature (3380 °C) and the ability to be drawn out into a fine wire.

In order to prevent premature failure through oxidation, oxygen must be removed from the enclosing glass bulb. Small lamps are evacuated, creating a vacuum in the bulb, but larger lamps are filled with argon, which reduces filament evaporation at high temperatures. The efficacy of filament lamps is relatively low but increases with the larger sizes. Colour rendering is generally very good but does depend on the glass finish of the lamp.

Tungsten-halogen lamps

Adding a halogen, such as iodine, to the enclosure prevents evaporation and allows the lamp to be run at a higher temperature. Colour rendering is very good with these lamps.

Halogen cycle

If a halogen gas is present in a lamp with a tungsten filament, the atoms of tungsten that are driven off the filament attach to halogen molecules instead of collecting on the lamp wall. They are eventually returned to the filament and separated. The tungsten is deposited on the filament. The halogen gas molecules are free to circulate again and available to intercept other tungsten atoms.

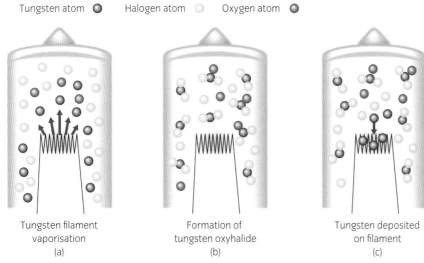

Tungsten atom ●   Halogen atom ○   Oxygen atom ●

Tungsten filament vaporisation
(a)

Formation of tungsten oxyhalide
(b)

Tungsten deposited on filament
(c)

Halogen regenerative cycle

Halogens, condense at about 300 °C, so the lamp must be kept above this temperature. The lamp is made of quartz glass, which can be weakened if touched by a person. Oil from the skin can lead to gas leaking from the glass. Handling these lamps without protection can therefore shorten their lifespan.

Tungsten-halogen lamps are widely used for floodlighting and vehicle-lighting applications as well as low voltage recessed lighting.

Discharge lamps

The way a heated filament produces light is relatively simple to understand. How light is emitted when an electric current flows in a gas or vapour can be more difficult to understand. Think about how in nature lightning produces large quantities of light when electric current passes through a gas (air).

Several types of lamp produce light by establishing a permanent electric arc in a gas. This process is known as electric discharge or gaseous discharge. It is used to produce light in fluorescent and high-intensity discharge lamps.

How discharge lamps work

Electrons are driven through the gas or vapour by the tube voltage, colliding with atoms as they go. The collisions are severe enough to break a loosely held electron from an atom, leaving behind a positively charged ion. This type of collision needs a fairly high tube voltage and results in ionisation to produce light.

Different gases and pressures of gas produce light of different wavelengths or colours, which can be further enhanced by using a coating such as phosphor around the inside of the gas tube.

When gas is cold, it has a high resistance and therefore requires a high-voltage 'strike' to ionise the gas. Once gas is ionising, its

ASSESSMENT GUIDANCE

Electronic starters prevent damage to the tube as they only allow a few attempts at starting, whereas a glow starter will continue to attempt to strike the tube.

resistance falls, meaning lower voltages can maintain ionisation. However, as proved by Ohm's law, a lower resistance will result in larger currents flowing. In order to create a high-voltage strike and limit running currents, discharge luminaires require control gear.

For a.c. supplies, an electronic control or an iron-cored inductor (also known as a choke or ballast) is used, rather than the resistor used for d.c. supplies.

To understand how the control gear works, look at the process, using a fluorescent luminaire (low-pressure mercury discharge lamp) as an example.

The sequence of events in this discharge luminaire is as follows.

1   When the luminaire is switched on, the starter switch closes and current flowing through the inductor induces a magnetic field within the inductor. As the gas in the tube is cold, the voltage is not enough to break down the resistance.

2   The starter switch opens, which open circuits the inductor. This causes the magnetic field in the inductor suddenly to collapse, creating a high voltage. This voltage strikes across the tube, ionising the gas, which reduces in resistance, and completing the circuit.

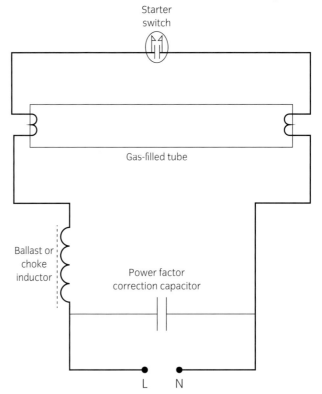

How control gear works in a discharge lamp

3   With the circuit once again complete, current flows through the inductor, which re-induces a magnetic field, causing self-induction, which limits the current flow. The starter switch is no longer required as the gas remains ionised with a constant current passing through it.

The switch starter

The switch starter is enclosed in a small glass tube containing neon gas, which glows and produces heat. The switch contacts are bimetals and the heat causes them to bend so that they touch.

When the fluorescent tube is lit, the current in the tube causes a voltage drop in the choke, so that the lamp voltage across the switch contacts is too low to cause a glow in the neon. As it is open circuit when unheated, the switch starter has no more effect on the luminaire circuit. If the tube fails to strike first time, the process repeats until striking is achieved.

Fluorescent tubes come in many lengths, shapes, colours and ratings.

Switch starter unit

Other types of discharge lamp

Other discharge lamps that work on a similar principle as the fluorescent tube (low-pressure mercury discharge lamp) above, include:

- high-pressure mercury
- low-pressure sodium
- high-pressure sodium
- metal halide.

High-pressure mercury (MBFU)

This type of lamp produces a near-white light with a blue tinge. It is commonly used for:

- amenity lighting
- street lighting in residential areas
- bollard lighting.

ACTIVITY

Look on the internet at a wholesaler's website to see the various types of fluorescent tube.

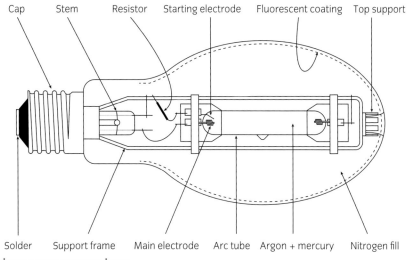

High-pressure mercury lamp

ASSESSMENT GUIDANCE

M = mercury

B = high pressure

SO = sodium

X = low pressure

N = high pressure

F = fluorescent coating

E = external ignitor

U = universal operation

Because of its good colour correction, it is good for CCTV applications or areas where coloured objects need identifying.

Low-pressure sodium (SOX)

As well as containing a low-pressure sodium gas, this type of lamp will also contain a neon-based gas that ionises at lower temperatures. The neon-based gas, which gives a pink appearance when starting, heats up the sodium gas, which then produces orange light.

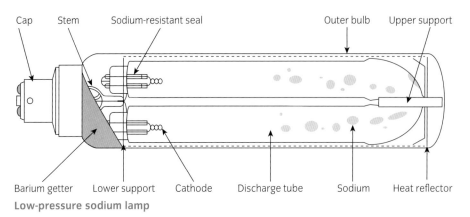

Cap   Stem   Sodium-resistant seal   Outer bulb   Upper support

Barium getter   Lower support   Cathode   Discharge tube   Sodium   Heat reflector

Low-pressure sodium lamp

These lamps have poor colour rendering but very good efficacy. They were widely used for most roadway applications but, due to increasing use of CCTV in towns and on roads, they are slowly being phased out and replaced with high-pressure sodium lamps.

High-pressure sodium (SON)

These lamps are commonly used for street and amenity lighting as well as car parks, high-bay lighting and security perimeter lighting. They have reasonably good colour rendering although the light output is light orange. They have a good efficacy despite the colour rendering, which is why they are a common choice of lamp. They come in two varieties: SON-E elliptical and SON-T tubular.

Some lamps contain internal ignition (starter) switches, but others rely on separate starter units within the control gear. If you are replacing one of these lamps, make sure you check the type needed, which should be indicated by the appropriate triangular symbol.

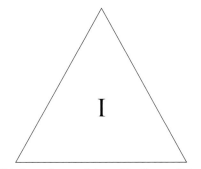

This lamp has an internal ignitor switch

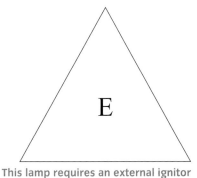

This lamp requires an external ignitor switch

High-pressure sodium lamp

Metal halide (HID)

Metal halide lamp

These lamps have excellent colour rendering and good efficacy. They are extensively used for sports arena floodlighting as well as general amenity or security lighting.

The Waste Electrical and Electronic Equipment Regulations 2006 (WEEE) and other environmental legislation place strict controls on the disposal of discharge tubes. Care is required when handling these tubes and lamps as the mercury in the tubes is toxic and the sodium in the lamps burns when in contact with moisture. It is therefore important that the tubes and lamps remain intact for specialist disposal.

Compact fluorescent lamps

These are miniature fluorescent tubes compacted into a small space. The control gear is contained in the base of the lamp. They are intended as energy-saving replacements for incandescent lamps although the colour rendering and flickering mean that many people find them difficult to use for reading or close work.

Compact fluorescent lamp

LED lights

The LED is a light-emitting diode. These lights are usually made from inorganic substances such as gallium indium nitride and gallium phosphide. The colour of the light output depends on the material used for the diode. The main colours are red, orange and green, and a variety of shades of blue.

The light output is usually monochromatic, ie the light emitted is at a single wavelength. The most common way to create a white light is to apply a phosphor-based coating to a blue diode. The phosphor converts the blue light to white light in a range of colour temperatures. The quality of the white light is affected both by the choice of LED and by the properties of the phosphor.

LEDs are very small; the active light-emitting surface is no bigger than 1–2 mm². A single diode can rarely produce enough light for a given lighting situation. For the unit to work, it must be mounted on a circuit board, with multiple LEDs in a cluster to form an LED module.

LEDs can be powered in two ways: with constant current or constant voltage. The ballast, which is referred to as the driver, is the unit that drives an LED array.

Basic LED colours

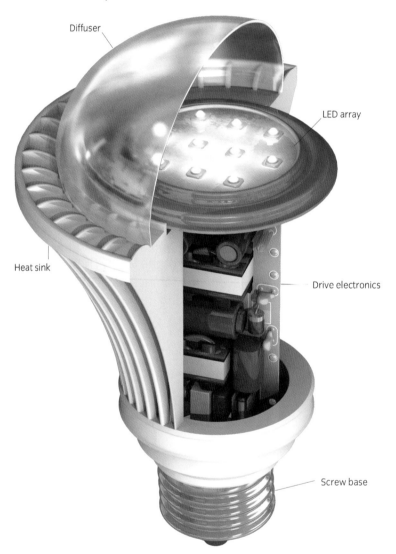

Retrofit LED lamp with cut-away showing internal parts

Although some LEDs run on conventional transformers, these can lack certain kinds of safety feature, such as short-circuit protection. When driven correctly, LEDs are claimed to be able to run for 50 000 hours, which is considerably longer than other technologies.

HOW ELECTRICAL SPACE-HEATING AND ELECTRICAL WATER HEATING SYSTEMS WORK

When current flows in a wire, apart from the flow of electrons, there is a thermal effect; the wire starts to heat up. The amount it heats up depends on factors such as the cross-sectional area of the wire, the amount of current flowing and the material that the wire is made of.

The heating effect of electricity is used in electrical heating systems such as electric fires, electric space heating and water heating.

Variations of this heat effect are also used to make light from light-bulb (lamp) filaments, which give off large amounts of light because they glow white hot as a result of the current passing through the thin filament.

Electrical space heating systems

The three ways to transfer heat from one medium to another are:

- convection
- conduction
- radiation.

Many heat sources use a combination of these methods.

Electrical water heating systems

There are many different types of electrically powered water heaters, including rod-type immersion heaters and instantaneous water heaters.

Assessment criteria

11.1 Explain the basic principles of electrical space heating and electrical water heating

11.2 Explain the operating principles, types, limitations and applications of electrical space and water heating appliances and components

SmartScreen Unit 309
PowerPoint 11 and Handout 11

Operating principles of electrical space heating systems

Convection

Hot air rises and colder air falls through the process of convection. A simple convection panel heater mounted on a wall uses this principle to move warm air around a room.

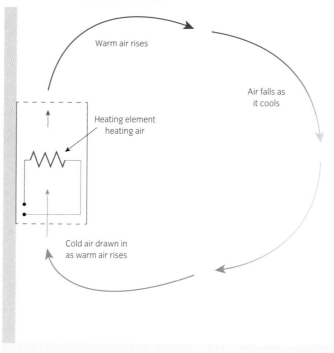

Warm air rises

Air falls as it cools

Heating element heating air

Cold air drawn in as warm air rises

Convection cycle

A convection heater usually has a low-temperature 'black-heat' element. Air in contact with the element is warmed and becomes less dense, so that it rises and is replaced by colder air, which is then warmed in turn.

Some convection heaters heat up another medium by conduction. For example, the element can be submersed in oil. The oil transfers heat around the unit, giving a larger body of heat to start the convection cycle.

Traditionally, convection heaters such as central heating emitters (radiators) are positioned where colder air is present, such as under windows, as this produces a larger cycle effect. This is less relevant today as modern windows provide better insulation.

Conduction

Conduction is the effect of heating something by direct contact. For example, the underfloor heating elements directly below a tiled floor warm the tiles and heat is transferred up through them. Similarly, in an immersion heater, a heating element is placed in water and the heat is transferred from the element directly to the water.

ASSESSMENT GUIDANCE

Immersion heaters are usually considered for back-up heating, perhaps in conjunction with solar thermal systems.

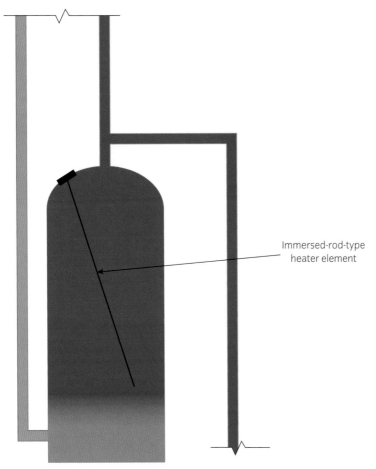

Immersion water heater heating water by conduction

Radiation

In this process heat is radiated (or thrown out) from a source and warms objects nearby. A standard coal fire does this; the heat can be felt on surfaces facing the fire, but not on surfaces facing away from the heat source.

Radiant heaters include:

- traditional electric fires
- infrared heaters
- log-burning stoves.

Heat sources

Underfloor heating

Underfloor heating systems have been around for centuries, although electrically powered versions have been available for a much shorter period. Since the 1960s the popularity of electric underfloor heating has fluctuated. In more recent times, it has become particularly associated with bathrooms and tiled areas. In many cases, underfloor heating is only installed in new houses or extensions as it is very costly to install into existing floors.

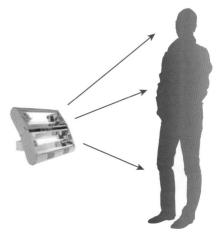

Radiant heat

Most heating devices use more than one of these methods of heat transfer.

Electric underfloor heating before the final screed is laid

Underfloor heating is an effective system of transferring heat into the floor surface by conduction and then heating the room by radiant heat and convection. It is a very good way of providing a uniform heat in a space.

Storage heaters

Storage heaters charge up at night when energy is available at low cost on off-peak tariffs. The energy is then stored in the unit in fireclay blocks, which release it slowly during the day. Radiator-type and fan-type storage heaters are available.

The radiator type is heated by elements in fireclay blocks. The release of heat is controlled by insulation. The storage heater is sized to store enough heat to last all day, under controlled release conditions.

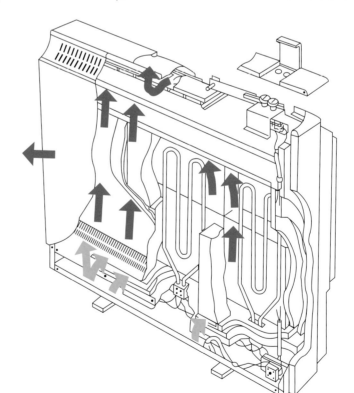

Electric storage heater

Fan-type storage heaters have thicker insulation so that very little heat is lost. A small unit, on a 24-hour supply for the fan, will provide warm air controlled by a thermostat. Because of the fan, it is possible to use up the heating charge, so there is often a short boost option. However, daytime boosting is not an economical way to use storage heaters.

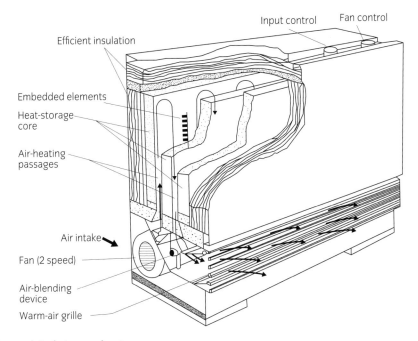

Input control   Fan control

Efficient insulation

Embedded elements

Heat-storage core

Air-heating passages

Air intake

Fan (2 speed)

Air-blending device

Warm-air grille

Fan-assisted storage heater

The fan-assisted storage heater is usually larger and always noisier than the radiator-type equivalent.

Panel heaters
Panel heaters heat spaces by convection and radiation. They can be slim in design and can even be fitted to the front of storage heaters to provide daytime heating if the stored charge has been lost.

Most panel heaters are provided with thermostatic and time controls. Although they can be fairly expensive to run, they are often chosen as a means of heating reasonably small locations or locations where other services, such as hot-water central-heating systems, are difficult to install.

Radiant or infrared heaters
These types of heater are particularly useful in large, cold areas where heating the air is difficult. Examples are garage workshops and warehouses. The heat radiated warms bodies but not the air around them. They would work just as effectively in a vacuum.

An infrared heater

Operating principles of electrical water heating systems

ACTIVITY

Why are infrared heaters commonly used in bathrooms?

Immersion heater systems

Immersion heater systems usually contain large amounts of water, which are heated over a period of time. The size of the vessel and limited amount of circulation and mixing allows relatively hot water to be used on demand. Then the heat is replenished over a period of time.

A large copper or stainless steel vessel is filled with water from a separate cold-water tank or sealed pressurisation unit. The vessel has an electric heater element fitted through a screw-threaded fitting, known as a boss.

The heater element can be the length of the cylinder, but in some instances two shorter elements are used, one positioned high and the other positioned low in the vessel. In domestic premises, the two heater elements used to be known as the 'bath and sink' function. By switching on the top element, just a small proportion of the water is heated, saving energy. This relies on the stratification process (ie hot liquids stay at the top) and works if the water is used before significant circulation takes place.

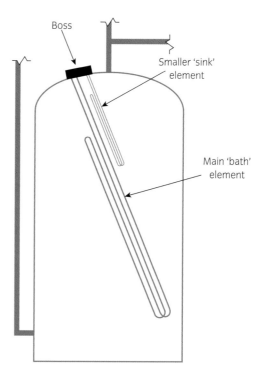

Cylinder element arrangements

Top entry boss with 'bath and sink' switch

To ensure that there is a full tank of hot water, a longer element is fitted to heat the whole tank. Temperature is controlled by a rod thermostat fitted in a pocket tube in the head of the heater. However, there are also strap-on thermostats that fix to the exterior of the hot water cylinder. Many immersion heaters also have a thermal cut-out to open the element circuit, should the thermostat fail and the water reach dangerously high temperatures, which could lead to enormous pressures in the tank or the venting of scalding water through the overflow or expansion pipe.

In order to save energy, thermal insulation is added to the vessel during manufacture or a thermally insulated jacket is fitted after installation.

Instantaneous water heaters

There are many examples of instantaneous water heaters; most common are electric showers and point-of-use hand washers, fitted above or below sinks.

With all of these heaters there is a limit to how much water can be raised to a specific temperature in a given time. This depends on the flow rate. The slower the flow rate, the hotter the water will get.

There are two types of instantaneous hot-water system. The tank type is like a miniature hot water cylinder. In the other type, the elements wrap around the water pipe inside the unit.

Thermostat inside pocket in heater element

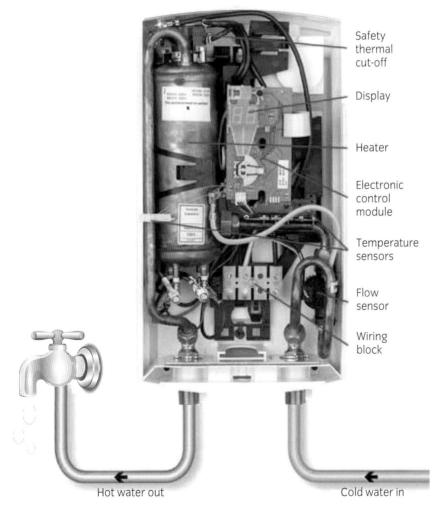

Safety thermal cut-off

Display

Heater

Electronic control module

Temperature sensors

Flow sensor

Wiring block

Hot water out   Cold water in

Over-sink tank-type instantaneous water heater

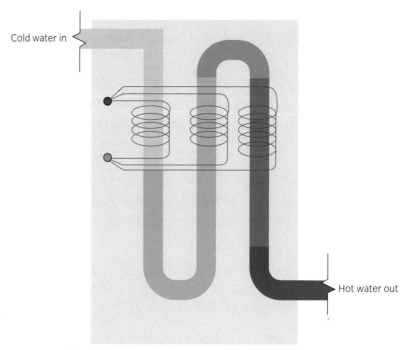

Elements around the water pipe in an instantaneous water heater

How to control heating systems

Heating and hot-water systems are not controlled just for economic reasons. Control is also a legal requirement for safety reasons. Building Regulations demand control.

Room thermostats and control circuits

Room thermostats are used to provide temperature control when heating spaces. Traditionally, this is done by means of a simple adjustable bimetallic sensor incorporating a set of contacts. When the desired temperature is reached, the contacts open due to the bimetal bending. This stops water-based heating systems pumping heat around or, on more advanced systems, operates a valve that shuts off the water, but still heats water until it reaches the desired temperature, when the pump or boiler/heater will switch off.

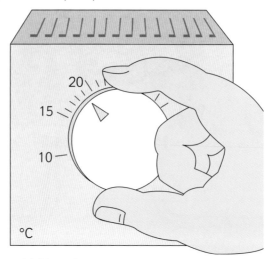

Simple thermostat with bimetal

The valve control arrangement is always used in commercial premises as it gives more accurate and zoned control, as required by Approved document L to the Building Regulations.

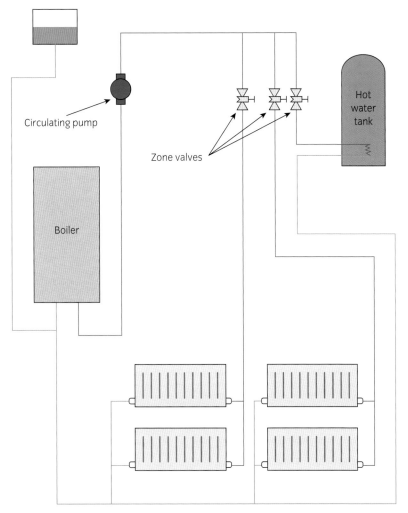

Simplified central heating system

Digital thermostat with temperature reading

As the bimetal thermostatic control is accurate to only plus or minus 3 Celsius degrees, many commercial and some domestic systems use digital thermostats containing temperature-sensitive electronic thermistors. These units give a signal that can be converted directly into a temperature, usually with an accuracy of up to 0.1 of a Celsius degree.

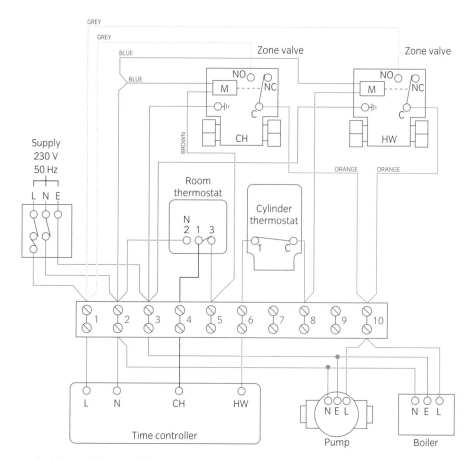

Typical domestic central heating circuit diagram

Simple daily time switch

Time switches and programmers

Time switches and similar devices are used to provide energy at the correct time and minimise wastage (eg not heating an office at weekends or a house when everyone is out).

In its simplest form, a 24-hour time switch cannot differentiate between days of the week. This is wasteful and a nuisance on days when the heating is required at different times from the norm. A programmable time controller is therefore normally installed as a minimum requirement for Building Regulations and convenience.

Simple domestic programmers allow the user to select the temperature required at set points on individual days of the week.

In commercial environments, controllers can be more sophisticated with optimum start functions. The user inputs the time at which they wish a specific temperature to be reached. The programmer uses thermostats to estimate the start-up time in order to reach the specified temperature and the time to switch off again in order to reduce the temperature at the end of the day.

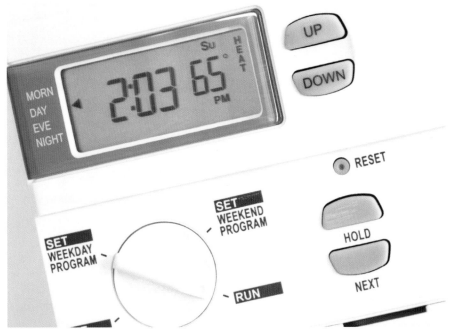

Programmer

ASSESSMENT GUIDANCE

Larger heating systems should be split into zones to allow accurate control.

Taking into account the thermal mass of the building, and the actual temperatures inside the building, the system adjusts itself for the next day. As with all systems that use computers to estimate comfort conditions, this type of system is often criticised because British weather is very unpredictable and conditions can change rapidly.

Understand the types, applications and limitations of electronic components in electrotechnical systems and equipment

Assessment criteria

12.1 Describe the function and application of electronic components that are used in electrotechnical systems

THE FUNCTION OF ELECTRONIC COMPONENTS IN ELECTRICAL SYSTEMS

Most electronic systems use many different electrical components in their power supply and in their operational systems.

Security alarm systems use full-wave rectification and smoothing through capacitors to supply a 12 V operating system. In addition, the closed-loop system uses transistor and similar technology to convert low-level signals from components such as passive infrared (PIR) detectors into an alarm output signal to components that operate the alarm.

Thyristors or SCRs are used extensively in motor speed-control circuits for heating and other applications, where motors and pumps require variable output. The ability to control an output waveform is essential in controlling the motor speed.

Heating control systems use a number of components. The most important element of any form of heating control is probably the ability to sense temperature in the airspace or water systems that are being heated.

A range of different detection devices using different technologies

A thermistor is used to determine accurately the temperature of the space or heating medium. The heating control system then uses feedback from the sensor to determine how much heat needs to be passed. The temperature is controlled via valves and/or variable speed pumps.

Diacs and triacs are used in lighting control circuits. The ability to trigger the device through a separate voltage allows dimming to be provided via proprietary dimming systems.

Typical thermistor-based space temperature detector

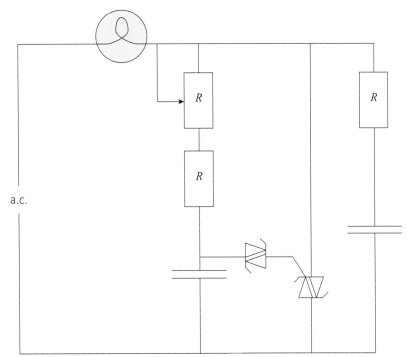

Typical diac and triac controlled dimming arrangement

HOW ELECTRONIC COMPONENTS WORK

In today's world of micro-components, it is becoming increasingly common to replace a whole circuit board rather than replacing single components. Nevertheless, it is far easier to diagnose faults in electrical systems if you understand how particular electronic components function.

Diodes

A diode is a silicon P-N junction, which allows current flow in one direction, but not the other. When current flows through a diode, it is called 'forward bias'. When current is restricted, it is called 'reverse bias'. There are several types of diode, from the simple one described above, used for rectification or signalling, to:

- a zener diode, which only allows current flow when a set voltage is reached

- a light-emitting diode (LED), which emits light when current flows through it

- a photo diode, which allows forward bias current flow when it detects light.

Symbols for different types of diode

SmartScreen Unit 309
PowerPoint 12 and Handout12

Assessment criteria

12.2 State the basic operating principles and applications of electronic components

KEY POINT

Always take care when handling or replacing electronic components or circuit boards as many can be damaged by static electricity. It is always important to ensure that static risks are minimised by earthing yourself to the equipment before handling components.

ACTIVITY

How could a capacitor, zener diode and resistor be used to smooth the output of a full-wave bridge rectifier?

Diacs

Symbol for a diac

A diac is a junction of two zener diodes, with two terminals. It works on a.c. circuits, hence the name **di**ode for **a.c.** A diac will not allow current flow unless a pre-set voltage is reached. Once this voltage is reached, current can flow in both directions. Current will continue to flow until the voltage falls below the level set, at which point the diac restricts current flow.

Thyristors

A thyristor is a solid-state switch that allows current flow between two of its terminals if a small current is sensed on the third. There are two types: silicon-controlled rectifiers (SCR) and triacs.

SCRs

The SCR is similar to the diode, in that current can only flow between the anode and cathode in one direction. However, it also has a gate terminal, which activates the switch when a small current is sensed on that terminal. Essentially, it allows a large current to be controlled by a small current. The SCR will continue to allow current flow between anode and cathode until the gate current is stopped. It does not require a constant gate terminal current, except when allowing the main current to pass.

Triacs

A triac has three terminals, one called the gate. If the gate senses a very small control current, a.c. is allowed to flow between the other two main terminals (known as MT_1 and MT_2). If the gate current is removed, the device will stop current flow when the alternating cycle reaches 0 V.

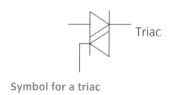

Symbol for a triac

Transistors

The transistor is the fundamental building block of modern electronics and the reason why electronic systems are now so affordable.

The three terminals on a bipolar transistor are known as base (B), collector (C) and emitter (E).

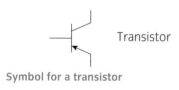

Symbol for a transistor

A transistor may be used either as a switch or an amplifier. When the base of an NPN transistor is grounded (0 V), no current flows between emitter and collector, so the transistor is off. If the base voltage is increased above 0.6 V, a current will flow from emitter to collector and the transistor is on. If the base current varies in value, the emitter to collector current will follow this pattern of variation with a larger or smaller current flow; in this situation, the transistor acts as an amplifier.

A PNP transistor operates in the same way as an NPN transistor but with current flow allowed in the reverse direction.

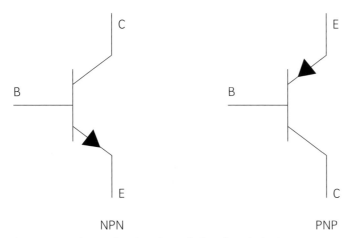

An NPN and a PNP transistor, showing the polarity of each device

Another type of transistor is the field effect transistor (FET), which has terminals marked gate, source and drain. The FET is much cheaper to produce as it requires less silicon. It also has the major advantage of operating at virtually no current on the gate terminal as long as a voltage above 0.6 V is present.

Resistors

Fixed resistor   Variable resistor   Light dependent resistor

Symbols for different types of resistor

Resistors are used to control or reduce current flow in electronic circuits. With a sufficiently high resistance, they can also be used as voltage dividers on certain circuits to allow a fixed voltage, less than the input voltage, to be obtained. Fixed-value resistors are either made from carbon film with an insulated coating, as shown below, or are wire wound for larger power applications.

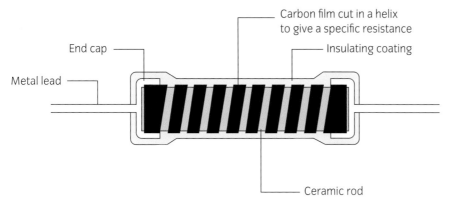

Carbon film cut in a helix to give a specific resistance

End cap

Insulating coating

Metal lead

Ceramic rod

Section through a carbon-film resistor

Carbon-film resistors are colour-coded to indicate their value and tolerance as shown below:

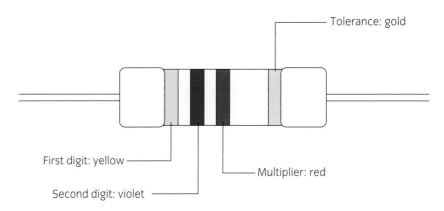

Resistor colour-coding system

Resistor colour values

Colour	Digit	Multiplier	Tolerance
Black	0	1	
Brown	1	10	1.0%
Red	2	100	2.0%
Orange	3	1 000	
Yellow	4	10 000	
Green	5	100 000	0.5%
Blue	6	1 000 000	0.25%
Violet	7	10 000 000	0.1%
Grey	8		0.05%
White	9		
Gold		0.10	5.0%
Silver		0.01	10.0%

KEY POINT

A resistor is identified by the colour bands red, violet, orange and gold. What is its value?

Wire-wound resistors are normally coded in order to establish the value, eg a 2R resistor is 2 Ω, whereas 2R2 is 2.2 Ω.

Thermistors

T   Thermistor

Symbol for a thermistor

A thermistor is a type of resistor in which the resistance varies significantly with temperature. This variation is so well defined that there is a definite temperature-related use for them. Thermistors typically achieve high precision within a limited temperature range, typically –90 °C to 130 °C.

Thermistors are widely used as temperature sensors, self-resetting overcurrent protectors for self-regulating heating elements and current inrush limiters.

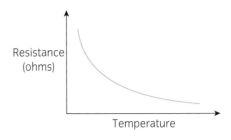

Effect of temperature on resistance in a thermistor

Photoresistors

These are resistors that vary in resistance, depending on the amount of light falling on them. They are often referred to as photocells and are used to control lighting as day/night switches.

Variable resistors or potentiometers

These resistors are used to vary resistance in a circuit manually. Their applications are wide, including use as sound volume controllers and speed controllers.

Capacitors

Capacitors are widely used in electrical circuits in many common electrical devices. A capacitor is a passive two-terminal electrical component used to store energy electrostatically in an electric field, rather than by chemical reaction, as in a battery. (Originally capacitors were known as condensers, but the original term has now been widely superceded.)

Capacitors vary widely, but all contain at least two electrical conductors separated by a dielectric (insulating layer), which acts as an insulator between the conducting plates. The plates are usually made from foils. The capacitance is varied by the area of the plates and the size of gap between the plates. The narrow gaps that are used require a very high dielectric strength.

Symbol for a capacitor

ACTIVITY

Name five different types of capacitor.

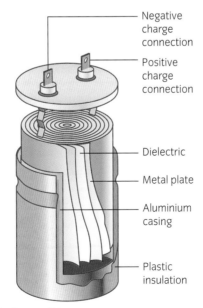

- Negative charge connection
- Positive charge connection
- Dielectric
- Metal plate
- Aluminium casing
- Plastic insulation

Construction of a typical capacitor

Rectifiers

A rectifier is an electrical device that uses diodes to convert alternating current (a.c.), which periodically reverses direction as it cycles, to direct current (d.c.), which flows in only one direction.

Half-wave and full-wave rectifiers are available.

Half-wave rectification

In half-wave rectification of a single-phase supply, either the positive or negative half of the a.c. wave is passed, while the other half is blocked. Half-wave rectification requires a single diode in a single-phase supply, or three in a three-phase supply.

As only one half of the input waveform reaches the output, the mean voltage is lower than full-wave rectification.

ASSESSMENT GUIDANCE

Three-phase rectification requires three diodes for half-wave and six diodes for full-wave rectification.

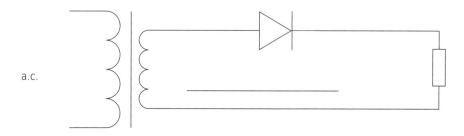

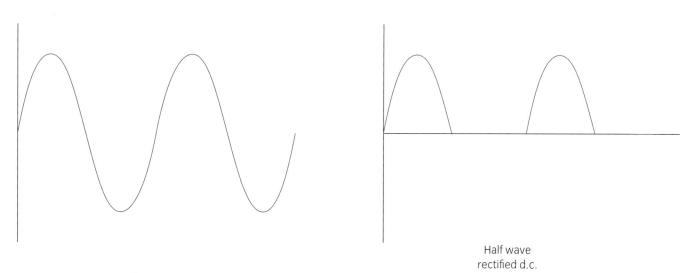

Input a.c.

Half wave rectified d.c.

Half-wave rectification

Full-wave rectification

A full-wave rectifier converts the whole of the input, both positive and negative components of the waveform, to one of constant polarity at its output. Full-wave rectification output gives a pulsating d.c waveform with a higher average output voltage than its half-wave counterpart.

The unit works with two diodes and a centre-tapped transformer, or four diodes in a bridge configuration, as shown below.

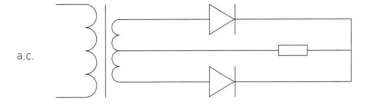

Two-diode method with centre-tapped transformer

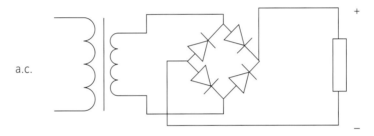

Four-diode or bridge rectifier

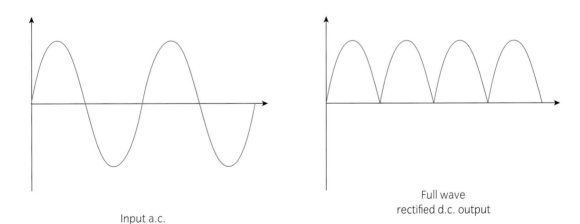

Input a.c.

Full wave rectified d.c. output

Full-wave rectification using two diode and bridge arrangements

ASSESSMENT CHECKLIST

WHAT YOU NOW KNOW/CAN DO

Learning outcome	Assessment criteria	Page number
1 Understand mathematical principles which are appropriate to electrical installation, maintenance and design work	*The learner can:* 1 Identify and apply appropriate mathematical principles which are relevant to electrotechnical work tasks.	2
2 Understand standard units of measurement used in electrical installation, maintenance and design work	*The learner can:* 1 Identify and use internationally recognised (SI) units of measurement for general variables	16
	2 Identify and determine values of basic SI units which apply specifically to electrical variables	17
	3 Identify appropriate electrical instruments for the measurement and calculation of different electrical values.	18
3 Understand basic mechanics and the relationship between force, work, energy and power	*The learner can:* 1 Specify what is meant by mass and weight	20
	2 Explain the principles of basic mechanics as they apply to levers, gears and pulleys	21
	3 Describe the main principles of the following and their inter-relationships: ■ force ■ work ■ energy (kinetic and potential) ■ power ■ efficiency	24
	4 Calculate values of electrical energy, power and efficiency.	27

6 Understand electrical supply and distribution systems

The learner can:

1 Describe how electricity is generated and transmitted for domestic and industrial/commercial consumption — 61

2 Specify the features and characteristics of a generation and transmission system — 63

3 Explain how electricity is generated from other sources — 66

4 Describe the main characteristics of: — 68
  - single-phase electrical supplies
  - three-phase electrical supplies
  - three-phase and neutral supplies
  - earth-fault loop path
  - star and delta connections

5 Describe the operating principles, applications and limitations of transformers — 76

6 State the different types of transformer that are used in electrical supply and distribution networks — 77

7 Determine by calculation and measurement: — 82
  - primary and secondary voltages
  - primary and secondary current
  - kVA rating of a transformer.

9 Understand the operating principles of different electrical components	*The learner can:*	
	1 Specify the main types and operating principles of the following electrical components:	129
	■ contactors	
	■ relays	
	■ solenoids	
	■ overcurrent protection devices:	
	□ fuses (HRC, cartridge and re-wireable)	
	□ circuit-breakers	
	■ RCDs	
	■ RCBOs	
	2 Describe how the following components are applied in electrical systems/equipment and state their limitations:	136
	■ contactors	
	■ relays	
	■ solenoids	
	■ overcurrent protection devices:	
	□ fuses (HRC, cartridge and re-wireable)	
	□ circuit-breakers	
	■ RCDs	
	■ RCBOs.	
10 Understand the principles and applications of electrical lighting systems	*The learner can:*	
	1 Explain the basic principles of illumination and state the applications of:	141
	■ inverse square law	
	■ cosine law	
	■ lumen method	
	2 Explain the operating principles, types, limitations and applications of luminaires.	148

11 Understand the principles and applications of electrical heating	*The learner can:*	
	1 Explain the basic principles of electrical space-heating and electrical water-heating	157
	2 Explain the operating principles, types, limitations and applications of electrical space- and water-heating appliances and components.	157
12 Understand the types, applications and limitations of electronic components in electrotechnical systems and equipment	*The learner can:*	
	1 Describe the function and application of electronic components that are used in electrotechnical systems	168
	2 State the basic operating principles and applications of electronic components.	169

ASSESSMENT GUIDANCE

Assessment A

- This is a closed book online e-volve multiple-choice assessment covering learning outcomes 1–4 of Unit 309.
- The paper has 30 questions and the time allowed is 1 hour.
- You should allow 2 minutes per question.
- Attempt all questions.
- Do not leave until you are confident that you have completed all questions.
- Keep an eye on the time as it moves quickly when you are concentrating.
- Make sure you read each question fully before answering.
- Ensure you know how the e-volve system works. Ask for a demonstration if you are not sure.
- Do not take any paperwork with you into the assessment.
- If you need paper to work anything out, ask the invigilator to provide some.
- Make sure your mobile phone is switched *off* (not on silent) during the assessment. You may be asked to give it to the invigilator.

Assessment B

■ This is a closed book short-answer assessment covering learning outcomes 5–12 of Unit 309.

■ The paper has 30 questions and the time allowed is 2 hours.

■ You should allow 4 minutes per question.

■ You will need drawing equipment such as a rule, protractor and pencils.

For Assessment A and Assessment B

■ You are allowed to use a scientific calculator (not programmable).

■ Keep your work neat and tidy. Don't lose marks due to poor workmanship making it difficult to mark.

■ Make sure you arrive in plenty of time for the assessment. Use all the time available – it is not a race to see who finishes first.

■ Go back over your work and check your answers.

Before the assessment

■ You will find some questions starting on page 183 to test your knowledge of the learning outcomes.

■ Make sure you go over these questions in your own time.

■ Spend time on revision in the run-up to the assessments.

OUTCOME KNOWLEDGE CHECK

1 For the following formula, what would be the correct answer if E is the subject?

$V = E - IR$

a) $E = V - IR$

b) $E = V + IR$

c) $E = IRV$

d) $E = \dfrac{IR}{V}$

2 The fraction $\dfrac{7}{16}$ expressed as a percentage would be

a) 0.437%

b) 7%

c) 16%

d) 43.7%.

3 The SI units of time and mass are respectively

a) minute and kg

b) second and pound (lb)

c) second and kg

d) minute and pound.

4 The unit of impedance is the

a) volt

b) watt

c) coulomb

d) ohm.

5 A mass of 10 kg is raised through 5 m. The energy required would be

a) 2 J

b) 50 J

c) 250 JJ

d) 490 J.

6 A motor has an input of 5.8 kW and an output of 4.75 kW. The efficiency of the motor would be

a) 1.05%

b) 10.5%

c) 82%

d) 122%.

7 An electric heater consumes 5.4 MJ in 45 minutes. The rating of the element would be

a) 1 kW

b) 1.5 kW

c) 2 kW

d) 3 kW.

8 Which one of the following is the best conductor of electricity?

a) Aluminium.

b) Copper.

c) Gold.

d) Silver.

9 Resistors of 8 Ω, 16 Ω and 24 Ω are connected in series. The total resistance will be

a) 12 Ω

b) 24 Ω

c) 48 Ω

d) 96 Ω.

10 Three resistors, each of 36 Ω are connected in parallel to a 48 V supply. The total resistance of the circuit will be

a) 12 Ω

b) 24 Ω

c) 36 Ω

d) 108 Ω.

KNOWLEDGE CHECK

1   Calculate the force on a conductor carrying 10 A in a field with a flux density of 0.8 tesla. The effective length of the conductor is 300 mm.

2   A circuit consists of a 2 V cell, a switch and a solenoid.

   a) Draw the circuit.

   Indicate where you would expect to measure:

   b) electromotive force

   c) electromagnetism.

3   A circuit consists of two resistors of 18 Ω and 22 Ω in series. The current through the 18 Ω resistor is 6 A. Calculate the supply voltage to the circuit.

4   Three resistors of 20 Ω, 10 Ω and 8 Ω respectively are connected in parallel to a 6 V d.c. supply. Calculate:

   a) the total resistance.

   b) the current taken from the supply.

5   The peak value of a sine wave value is 350 V. Calculate the

   a) RMS value

   b) average value of a half cycle

   c) peak to peak value.

6   An a.c. generator running at 15 r/s produces an RMS output of 240 V at 60 Hz. If the speed is increased to 20 r/s, state the effect on

   a) voltage output

   b) frequency

   c) periodic time.

7   Draw the circuit diagram of a d.c. series motor and identify the two main components.

8   Show how the following loads may be connected to a 400 V four-wire supply.

   a) 230 V single-line supply

   b) 400 V three-line and neutral supply

   c) 400 V single-line supply.

9   Three single-phase loads of

   a) 60 A, b) 50 A and c) 40 A are connected to a three-line supply, all at unity power factor. By any method, determine the value of the neutral current.

10 Identify three 'green' sources of electricity used in the United Kingdom.

11 State the operating principle of a single-phase double-wound transformer.

12 Sketch the layout of a three-phase, shell-type transformer, including the position of the primary and secondary windings.

13 A 200 V a.c. circuit consists of a coil of 30 Ω reactance and resistor of 50 Ω. A capacitor of 30 Ω reactance is connected in series with the coil.

Calculate the

a) impedance

b) current

c) power.

14 A circuit consists of a capacitor connected in series with a resistor. Draw an impedance triangle indicating the location of resistance, capacitive reactance and impedance.

15 A fluorescent lighting scheme is to be replaced by equivalent-power new fittings. The electrician mistakenly orders low power factor fittings instead of those fitted with capacitors.

State:

a) the effect on the supply current if LPF fittings are used.

b) the position in which capacitors should be fitted.

16 Identify three advantages of using cartridge fuses rather than rewireable fuses in an installation.

17 A load of 2.5 kW takes a current of 12 A from a 230 V supply. Calculate the power factor.

18 Explain how the split-phase effect is achieved in a capacitor-start induction-run single-phase motor.

19 Explain the basic principle of an immersion-type heater.

20 State, for each of the following components, one typical application:

a) LED

b) triac

c) transistor.

GLOSSARY

A

Algebra The branch of mathematics that uses letters and symbols to represent numbers, to express rules and formulae in general terms.

C

Closed circuit A complete circuit connected to a source of energy. If the circuit contains a switch and the switch is switched off, it becomes an open circuit.

Common denominator A denominator that can be divided exactly by all of the denominators in the question.

E

Earth fault loop impedance The impedance of the earth fault current loop starting and ending at the point of earth fault. This impedance is denoted by the symbol Z_s.

Efficacy The ratio of power in and power out, measured in two different units. For example, the ratio of light output in lumens to the electrical power measured in watts.

Efficiency The ratio between power in and power out, measured in the same unit.

Electrolyte A chemical solution that contains many ions. Examples include salty water and lemon juice. In major battery production, these may be alkaline or acid solutions, or gels.

F

Fulcrum The pivot point.

H

Hypotenuse The longest side of a right-angled triangle, which is opposite the right angle.

I

Induction motors The induction motor is the simplest form of alternating current (a.c.) motor. It is also known as the asynchronous motor.

J

Joule The unit of measurement for energy (W), defined as the capacity to do work over a period of time.

M

Mean The mean is the average of a set of numbers; a calculated 'central' value of a set of numbers. To calculate the mean, just add up all the numbers of the set, then divide by how many numbers there are in the set.

R

Root mean square (RMS) The square root of the mean of the squares of the value.

Rotor speed The actual speed at which the rotor rotates in revolutions/second (r/s)

S

SI units The units of measurement adopted for international use by the *Système International d'Unités*.

Slip The difference between the synchronous speed and the rotor speed expressed as a percentage or per unit value

Standard gravity Standard gravity is the average gravity on Earth but the actual value can alter depending on where you are.

Step-down transformer A transformer that has a higher number of turns on the primary than on the secondary.

Step-up transformer A transformer that has a higher number of turns on the secondary (output stage) than on the primary (input stage).

Synchronous speed The speed at which the field rotates around the stator field poles in an induction motor.

T

Torque A force that causes rotation.

Transposition Rearranging a formula to make the unknown you need to find, the subject of the formula.

ANSWERS TO ACTIVITIES AND KNOWLEDGE CHECKS

Answers to activities and knowledge checks are given below. Where answers are not given it is because they reflect individual learner responses.

UNIT 309 UNDERSTANDING THE ELECTRICAL PRINCIPLES ASSOCIATED WITH THE DESIGN, BUILDING, INSTALLATION AND MAINTENANCE OF ELECTRICAL EQUIPMENT AND SYSTEMS

Activity answers

Page

6   $V = E + I_a R_a$

$$N = \frac{E}{\phi}$$

$$I = \sqrt{\frac{P}{R}}$$

$$N = \frac{EA}{2P\phi Z}$$

$$\cos\phi \frac{P}{VI}$$

8   It is the order which mathematical operations are carried out and stands for Brackets, Orders, Division, Multiplication, Addition, Subtraction. (O may also stand for 'of' instead of 'orders'. In each case it means 'the power of'.)

15   £300.99.

17   a) U = voltage between the lines

     b) U_o = nominal a.c. rms voltage to earth

     c) U_{oc} = open circuit voltage.

18   A voltmeter has a very high resistance.

23   First.

24   Mainly friction losses in the bearings and the ropes passing over the pulleys.

27   $4 \times \dfrac{75}{100} \times \dfrac{55}{100} = 1.65$ kW

28   a) Carbon has 6 electrons. b) Aluminium has 13. c) Silicon has 12. d) Gold has 79.

31   The duration of a second is defined as 9 192 631 770 periods of the radiation

Page

     corresponding to the transition between the two hyperfine levels of the ground state of the caesium 133 atom.

36   They were converted from the old imperial sizes.

39   896 kg.

40   Silver, copper, gold, aluminium, brass, steel. As steel is an alloy, its resistance is dependent on the metal mix.

41   Table 4B1.

45   $\dfrac{3000}{230} = 13.04$ A

     $\dfrac{230}{13.04} = 17.66\,\Omega$

47   Sintered metal oxide (thermistors).

54   For example, wind generators, car alternators.

59   Examples of different supply frequencies are: Canada 60 Hz, France 50 Hz, USA 60 Hz, Spain 50 Hz.

60   a) 155 V; b) 339 V; c) 565 V.

63   E.ON, npower, SSE, British Gas, eDF Energy.

66   These tend to be hilly areas with deep valleys that can be flooded to store large quantities of water.

70   Three-phase motor, three-phase heater and so on.

71   To hold it at zero potential.

72   a)   230 V, b) 240 V, c) 6.351 kV, d) 63.5 V.

75   Main earthing conductor.

77   Autotransformer. A common winding is shared between the primary and secondary. In the step-up type the secondary has more turns and a higher voltage. In the step-down variety the secondary has fewer turns. The primary and secondary currents share a common winding.

78   Iron losses, which consist of eddy current and hysteresis loss.

78   3 A.

80   Toroid core.

87   Plot 0° = 0; 30° = 0.5 × 100 = 50 V;

      60° = 0.66 × 100 = 86.6 V;

      90° = 1 × 100 = 100 V; back to zero at 180° and then minus values back to 360°.

88   Candidates to plot power wave.

92   0.722 A.

95    $I_s = \dfrac{230}{115} = 2$ A; $V_{cap} = 2 \times 200 = 400$ V.

98   It costs more than is saved; 0.95–0.97 is the required range.

100   Refrigerator, vacuum cleaner, washing machine, dishwasher, tumble dryer or any other motor-driven appliance.

101   cos 1° = 0.9998, cos 10° = 0.98,

      cos 30° = 0.87, cos 45° = 0.71, cos 70° = 0.34,

      cos 90° = 1, cos^{-1} 0.75 = 41.41°

104   Power factor from phasor is approx 0.7 leading, 38.46°

105   Individual, advantage: automatically switched on and off with load; individual, disadvantage: may need more capacitance than bulk correction. Bulk, advantage: less capacitance than individual capacitors; bulk, disadvantage: only corrects up to point of installation.

110   a) 230 V, b) 240 V, c) 6.351 kV, d) 63.5 V.

113   Copper loss in fields; copper loss in armature; windage (rotational) loss; bearing loss; brush voltage drop.

116   The field of a self-excited machine obtains its supply from the armature output. A separately excited machine obtains the field supply from an independent source.

120   $N = \dfrac{f}{p}$, a) $N = \dfrac{40}{2} = 20$ rev / s,

      b) $N = \dfrac{50}{2} = 25$ rev / s,

      c) $N = \dfrac{60}{2} = 30$ rev / s.

122   Clocks, timers, compact disc player motors or any other suitable application where precise motor speed is important.

123   The electrolytic is used for starting; the paper capacitor stays in series with the run winding.

127   One-third.

135   BS 7671 states that the residual current rating must not exceed 30 mA.

137   Thermal and magnetic devices.

138   30 A – red, 45 A – green.

142   a) 200 lux, b) 50 lux.

160   Various names are used but 'Off-peak tariff' and 'Economy 7' are typical; any tariff that provides cheaper electricity during the night.

162   For short-term use; they heat the object (body) rather than the air.

169   Zener diode connected across output; when the breakdown voltage is reached it conducts and causes voltage drop across the resistor; the capacitor charges and holds the output voltage steady (or similar description).

170   It would vary the current flow through the device. Trigger at 90° and only the second half of the first half wave is conducted.

173   Any from mica, paper, electrolytic, air, tantalum, film, ceramic, vacuum.

Outcome knowledge check answers

Page 183

| 1 | b) | 2 | d) | 3 | c) | 4 | d) | 5 | d) |
| 6 | c) | 7 | c) | 8 | d) | 9 | c) | 10 | a) |

Knowledge check answers

1   $F = \beta LI$

   $F = 0.8 \times 0.3 \times 10$

   $F = 2.4\ N$

2   a)   circuit diagram

   b)   internal to cell

   c)   around coil.

3   Total resistance $= 18 + 22 = 40\ \Omega$

   Current is common in all parts of circuit

   $= I \times R = 6 \times 40 = 240\ V$

4   a)   By any suitable method,

   $R_t = 3.64\ \Omega$

   b) $I = \dfrac{V}{R} = \dfrac{6}{3.64} = 1.648A$

5   a) $350 \times 0.707 = 247.5$

   b) $350 \times 0.636 = 222.6$

   c) $350 \times 2 = 700\ V$

6   a)   Voltage will increase

   b)   Frequency will increase

   c)   Periodic time will decrease.

7

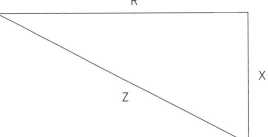

field   armature

8   a)   between any line and neutral

   b)   connect to all three lines and neutral between any two lines

   c)   between any two lines.

9   16.5 A.

10   Any from: wind power, wave power, solar power (PV), micro hyro.

11   The current flowing in the primary produces a primary magnetic flux. Which circulates within the transformer core. This flux cuts the secondary coil and induces an emf in it (Mutual Induction).

12   a)   suitable sketch of three limb core with

   b)   both primary and secondary windings on centre limb.

13   a)   X will now be $(30–30) = 0\ \Omega$ and Z will be the same as R.

   b) $I = \dfrac{V}{Z} = \dfrac{200}{50}\quad I = 4\ A$

   c) $P = V \times I \times \cos\theta$

   $P = 200 \times 4 \times 1 = 800\ W$

14

R

X

Z

15   a)   The current would increase.

   b)   The capacitors should be fitted across the supply terminals of the fluorescent.

16   1.   Cartridge fuses have a higher breaking capacity than rewireable fuses.

   Rewireable S4 4 kA

   BS 1361 Type II 33 kA

   BS 88 up to 80 kA.

   2.   Cartridge fuses should not be replaceable with incorrect size.

3. Cartridge fuse has indicator to show when blown, rewireable has to be removed to check.

4. Cartridge fuse element not exposed to air so does not oxidise like rewireable fuse.

   Or suitable alternatives.

17 Power factor $= \dfrac{W}{VA}$

    P.F. $= \dfrac{2500}{230 \times 12}$

18 The start winding is made of a smaller diameter wire which gives it a higher resistance than the run winding. The angle between the start and run winding is quite small. To increase the angle a capacitor is connected in series with the run winding.

    Or simply, there is a phase shift which makes the motor start.

19 An insulated element is immersed in a tank of water. When the element heats up the heat produced is conducted to the water.

20 a) light emitting diode used as an indicator or more recently a general lighting source

    b) an a.c. switching device for lighting and motor control

    c) used as an amplifier or switching device.

INDEX